The ESSENTIALS of

Anatomy & Physiology

Jay M. Templin, Ed.D.
Professor, Biology Department
LaSalle University
Philadelphia, PA

Research & Education Association
61 Ethel Road West
Piscataway, New Jersey 08854

THE ESSENTIALS®
OF ANATOMY & PHYSIOLOGY

Printed in the United States of America

Library of Congress Catalog Card Number 98-66628

International Standard Book Number 0-87891-922-8

III-1

WHAT "THE ESSENTIALS" WILL DO FOR YOU

This book is a review and study guide. It is comprehensive and it is concise.

It helps in preparing for exams, in doing homework, and remains a handy reference source at all times.

It condenses the vast amount of detail characteristic of the subject matter and summarizes the **essentials** of the field.

It will thus save hours of study and preparation time.

The book provides quick access to the important principles, concepts, practices, and definitions in the field.

Materials needed for exams can be reviewed in summary form – eliminating the need to read and re-read many pages of textbook and class notes. The summaries will even tend to bring detail to mind that had been previously read or noted.

This "ESSENTIALS" book has been prepared by an expert in the field, and has been carefully reviewed to assure accuracy and maximum usefulness.

Dr. Max Fogiel
Program Director

CONTENTS

Chapter 6
THE SKELETAL SYSTEM

Chapter 7
THE SKELETAL MUSCLES

Chapter 8
THE NERVOUS SYSTEM

Chapter 9
THE SENSE ORGANS

Chapter 10
THE ENDOCRINE SYSTEM

CHAPTER 1

Introducing the Human Body

1.1 Anatomy and Physiology

Anatomy - Anatomy is the study of the structure of body parts. It is also the study of the relationship among these parts. The heart, for example, consists of chambers, valves, and associated blood vessels.

Physiology - Physiology is the study of the function of body parts. The parts of the heart, for example, work together to pump the blood throughout the body.

There is a close association between anatomy and physiology. Structure complements function. The four chambers of the heart have muscular walls that contract to pump the blood. The makeup of the valves prevents the backflow of blood.

1.2 Levels of Organization

The anatomy of the human body is composed of different levels of organization. These levels represent a series of steps. Each level is a building step for the next level.

These levels are:

Atom - All matter consists of elements. These simple substances exist as discrete, submicroscopic particles called atoms. The four most common elements of the human body are carbon, hydrogen, oxygen, and nitrogen.

Molecule - Atoms bond into molecules. About 65% of human body weight consists of water molecules. Smaller molecules bond into larger molecules that have biological functions. Monosaccharides (i.e., glucose), for example, bond into polysaccharides (i.e., starch). These carbohydrates are an energy source.

Organelle - Molecules compose the parts of the cell called organelles. Each of these parts carries out a specific function. The ribosome, for example, is the site of protein synthesis.

Cell - The cell is the smallest unit displaying the properties of life. Cells tend to specialize. There are about 200 different kinds of specialized cells in the human body. Neurons (nerve cells) send signals. Leukocytes (white blood cells) fight infection.

Tissue - Similar cells function together in a tissue. Muscle cells work together in skeletal muscle tissue. These cells contract, producing body movement.

Organ - Two or more tissues work together in an organ. The heart consists of several tissue types.

Organ Systems - Organs with related functions are part of the same organ system. The heart and blood vessels are organs of the circulatory system. They function to circulate the blood throughout the body.

Organism - All organ systems make up the organism. The organ systems of the human body include the nervous, circulatory, respiratory, and digestive systems.

1.3 Anatomical Terms

Anatomical terms are used to describe the makeup of the body accurately and concisely. All of these terms are used with reference to anatomical position. In this position the subject studied is facing forward and standing erect. The arms are hanging at the sides. The palms and toes are pointed forward.

1.3.1 Directional Terms

Directional terms compare the relative position of one body part to another body part. These terms occur in pairs. The members of each pair have opposite meanings.

Superior/Inferior - Superior means closer to the head. Inferior means closer to the feet. The neck is superior when compared to the chest, which is inferior. When compared to the abdomen, the chest is superior and the abdomen is inferior.

Anterior/Posterior - Anterior (ventral) refers to a part that is closer to the front of the body. Posterior (dorsal) refers to a part that is closer to the back. The heart is anterior when compared to the vertebral column, which is posterior. When compared to the sternum (breastbone), the heart is posterior and the sternum is anterior.

Medial/Lateral - Medial refers to a part that is closer to an imaginary midline passing vertically through the body. Lateral refers to a part that is further from this midline. The nose is medial when compared to the eyes (lateral). When compared to the ears, the eyes are medial. The ears are lateral.

Proximal/Distal - Proximal refers to a part of a limb that is closer to the trunk (torso) of the body. Distal refers to a limb part that is further from the trunk. The forearm is proximal when compared to the wrist (distal). The wrist is proximal when compared to the fingers (distal).

Other directional terms include:

Superficial - closer to the surface of the body
Deep - further away from the surface of the body
Parietal - referring to the wall of a body cavity
Visceral - referring to an organ within the body cavity

1.3.2 Planes and Sections of the Body

Incisions can be made through the body to study the internal anatomy. These sections represent imaginary planes.

Sagittal Plane - A sagittal plane passes through the body longitudinally, dividing it into left and right regions. A midsagittal section passes through the midline of the body.

Coronal (Frontal) Plane - A coronal plane passes through the body longitudinally, dividing it into anterior and posterior regions.

Transverse Plane - A transverse plane passes through the body horizontally, dividing it into superior and inferior regions.

These sections can also pertain to organs of the body. Sagittal, coronal, and transverse planes also pass through the heart.

1.3.3 Body Cavities

There are two main cavities of the human body, the dorsal cavity and ventral cavity. Each cavity is divided into subcavities.

Dorsal Cavity - The dorsal cavity consists of the **cranial cavity** and **spinal cavity**. The cranial cavity is formed by the superior bones of the skull. It contains the brain. The spinal cavity is formed by a series of vertebrae. It contains the spinal cord.

Ventral Cavity - The ventral cavity consists of **thoracic** and **abdominopelvic subcavities.**

The thoracic cavity is superior to the diaphragm. It is subdivided into a left and right **pleural cavity.** The pleural cavities contain the lungs. The **mediastinum** is the space between the pleural cavities. It contains the trachea (windpipe), esophagus, thymus gland, and heart. The heart is contained within a separate cavity of the mediastinum, the **pericardial cavity**.

The abdominopelvic cavity is inferior to the diaphragm. The larger abdominal portion contains the liver, gallbladder, stomach, small intestine, and most of the large intestine. The smaller pelvic portion contains the large intestine, bladder, and reproductive organs.

1.4 Organ Systems

Organs with related functions are part of the same organ system. These systems are:

Integumentary System - The skin and accessory organs comprise the integumentary system. This system protects and regulates body temperature.

Skeletal System - The skeletal system consists of the bones and articulations (joints). This system provides protection and support. Skeletal muscles pull on bones to produce movements. The skeletal system also stores minerals and produces blood cells.

Muscular System - The skeletal muscles contract to produce body movements. The muscles also produce body heat.

Nervous System - The nervous system sends signals throughout the body. The central nervous system consists of the brain and spinal cord. The peripheral nervous system consists of the cranial and spinal nerves.

Endocrine System - The glands of the endocrine system secrete chemical messages called hormones. These messages regulate processes such as growth and mineral balance.

Circulatory System - The circulatory system transports substances to and from body cells. The lymphatic system is part of the circulatory system. One of its functions is to protect the body from disease.

Respiratory System - The respiratory system distributes and exchanges gases between the body and external environment.

Digestive System - The digestive system prepares food molecules for use by the cells of the body.

Urinary System - The urinary system controls the composition and volume of the blood. It eliminates wastes.

Reproductive System - The male reproductive system produces sex cells. It transfers these cells to the female reproductive system. The female reproductive system produces sex cells and receives the male sex cells. It also provides the internal environment for the development of the embryo and fetus.

1.5 Homeostasis

Homeostasis is the maintenance of relatively constant conditions of the internal environment of the body. The internal environment includes the tissue fluid that bathes the cells. This tissue fluid is formed from the blood. Characteristics that are controlled include:

Temperature - at 37° C
Blood Sugar - 100 mg per 100 ml of blood
pH of the Blood - at 7.4

These characteristics are controlled by **negative feedback**. Receptors sense changes in the internal environment. Through signals from the nervous and endocrine systems, responses reverse the trend of these changes. For example, if blood sugar increases, the pancreas secretes insulin. This hormone signals responses that decrease the level of glucose in the blood.

If body temperature decreases, responses increase body temperature. One response is the shivering of the skeletal muscles.

CHAPTER 2

Chemistry of Life

2.1 Elements and Atoms

Element - An element is a substance that cannot be broken down into simpler substances by normal chemical reactions. The four most common elements of the human body are carbon, hydrogen, oxygen, and nitrogen. Elements exist as discrete, submicroscopic particles called atoms.

Atom - Each element consists of one kind of atom. An atom is the smallest particle that displays the chemical properties of an element. The three main types of subatomic particles of an atom are the **proton, neutron**, and **electron.**

Atomic Weight - The atomic weight (mass) is the total number of protons and neutrons in the nucleus of an atom. For example, the atomic weight of carbon is usually 12. For oxygen it is usually 16.

Atomic Number - The atomic number is equal to either the number of protons or electrons in the atom. For example, the atomic number of carbon is 6. For oxygen it is 8.

Isotope - Isotopes are atoms of the same element that have a different number of neutrons. Isotopes have different atomic weights. Oxygen is one atom having isotopes with the atomic weights of 16 (8 protons, 8 neutrons) and 18 (8 protons, 10 neutrons).

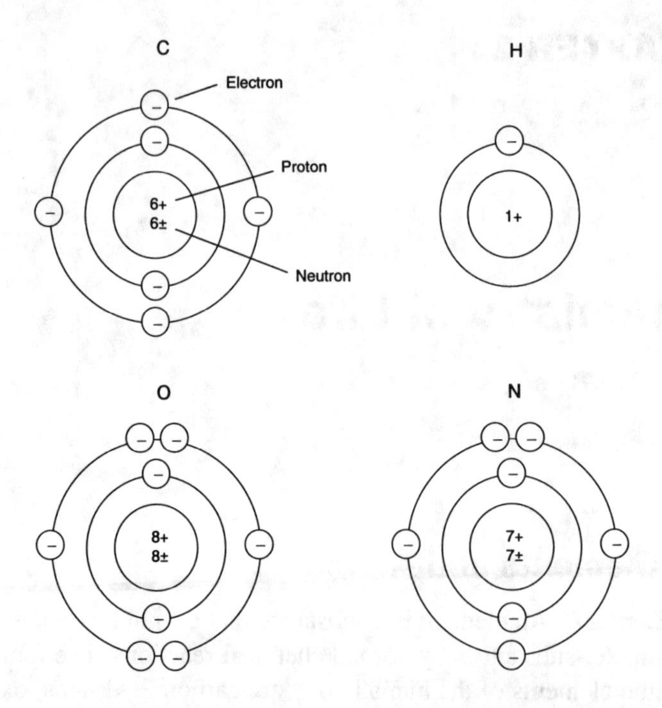

Figure 2.1 Planetary Models of Carbon, Hydrogen, Oxygen, and Nitrogen

Ions - An ion is an atom which has lost or gained electrons. If the sodium atom loses its one outershell electron, it becomes a positive ion (cation). If the chlorine atom gains one electron in its outer shell, it becomes negative (anion).

2.2 Water

About 65% of human body weight is water. A water molecule consists of two hydrogen atoms linked to one oxygen atom by polar covalent bonds. Its most important biological function is serving as a solvent for many kinds of solutes. A solution is a mixture of a solute and solvent.

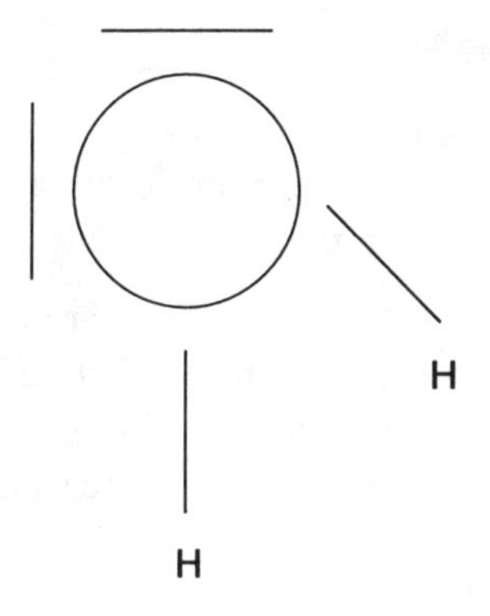

Figure 2.2 Structural Formula of a Water Molecule

Acid - An acid is a compound that dissociates in water to yield hydrogen ions (H+). An acid is a proton donor.

Base - A base is a compound that dissociates in water to yield hydroxyl ions (–OH). A base is a proton acceptor.

pH - pH is a scale showing the degree of acidity or alkalinity of a solution. The scale has a range of 0 to 14. A pH of 7 is neutral. A pH of less than 7 is acidic. A pH of more than 7 is alkaline (basic). The skin, for example, has a pH of about 5. The pH of the blood is about 7.4.

Buffer - Buffers are compounds that stabilize the pH of a solution. They react to resist changes in pH if an acid or base is added to a solution.

Salt - A salt is a compound consisting of a cation other than hydrogen and an anion other than the hydroxyl ion. One example of a salt is sodium chloride (NaCl). Salts dissociate into positive and negative ions (cations and anions) in water.

2.3 Organic Compounds

Organic compounds are complex compounds of carbon. They always include hydrogen. Other elements may be found in the molecules. There are four families of organic compounds with important biological functions.

2.3.1 Carbohydrates

Carbohydrate molecules consist of carbon, hydrogen, and oxygen. They have a general molecular formula of CH_2O. There are several subfamilies based on molecular size.

Monosaccharides - Monosaccharides are the building blocks of larger carbohydrate molecules. They are simple sugars. Important monosaccharides include **glucose** (blood sugar), **fructose**, and **galactose**. These three monosaccharides have the same molecular formula of $C_6H_{12}O_6$ but different structural formulas. Molecules with this relationship are called **isomers**.

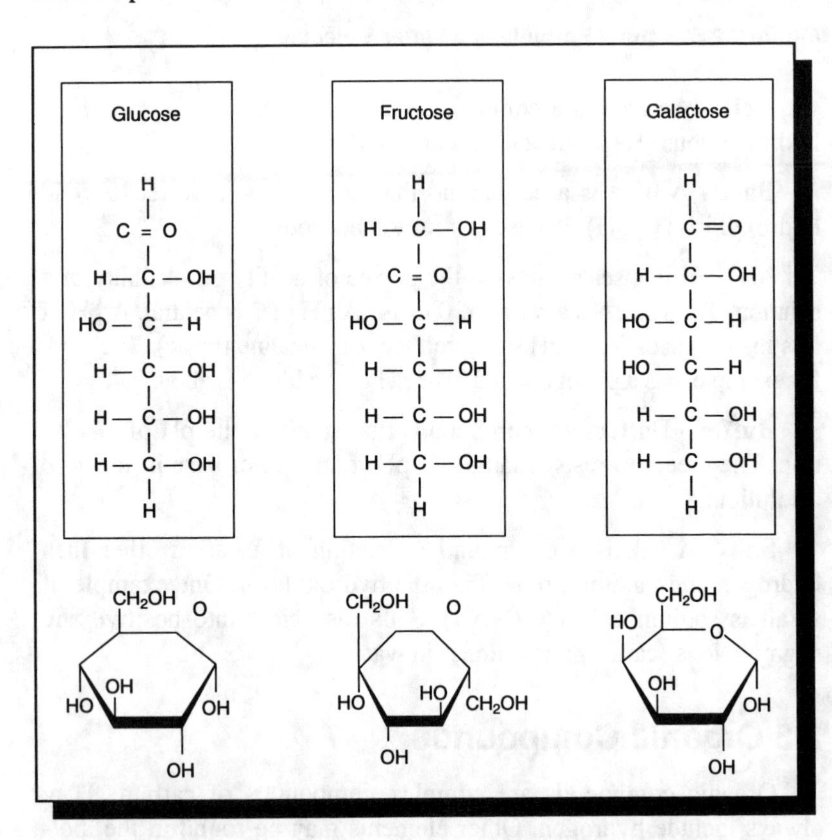

Figure 2.3 Structural Formulas of Glucose, Fructose, and Galactose

Disaccharides - Two monosaccharides bond together by a dehydration synthesis to produce a larger molecule, the disaccharide. By this process a molecule is lost between the two smaller molecules as they bond together, forming the larger molecule. Important disaccharides are **maltose** (glucose + glucose), **sucrose** (glucose + fructose), and **lactose** (glucose + galactose).

Polysaccharides - Many monosaccharides bond into long, chainlike molecules called polysaccharides. **Glycogen** is the main polysaccharide in the human body. It consists of many bonded glucose molecules. **Glycogen** stores energy. Glucose is produced when starch is broken down. Glucose is found in the blood. It offers an immediate source of energy to the cells of the body.

2.3.2 Lipids

Lipid molecules contain at least carbon, hydrogen, and oxygen. Some contain nitrogen and phosphorous. Lipids are insoluble in water (hydrophobic).

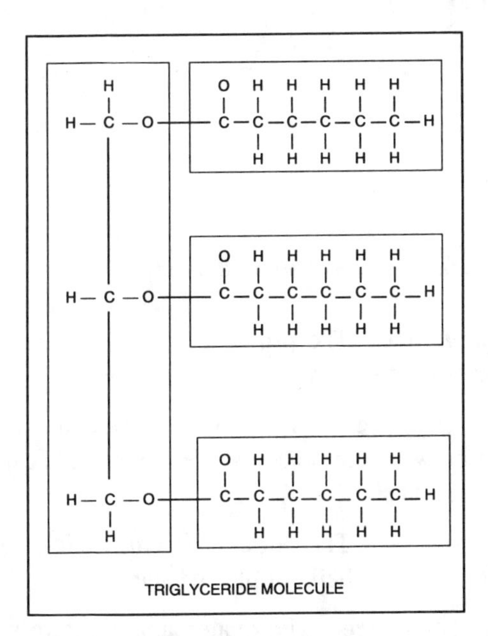

Figure 2.4 Structural Formula of a Triglyceride

Triglycerides - Triglycerides consist of three **fatty acids** bonded to a molecule of **glycerol**. They store large amounts of energy (9 calories/gram).

Phospholipids - Phospholipids are formed when one of the fatty acids of the triglyceride is replaced by a phosphate group. They are a major component of cell membranes.

2.3.3 Proteins

All proteins contain carbon, hydrogen, oxygen, and nitrogen. Some also contain phosphorous and sulfur. The building blocks of proteins are **amino acids**. There are 20 different kinds of amino acids used by the human body. They unite by peptide bonds to form long molecules called polypeptides. Polypeptides are assembled into proteins. Proteins have four levels of structure.

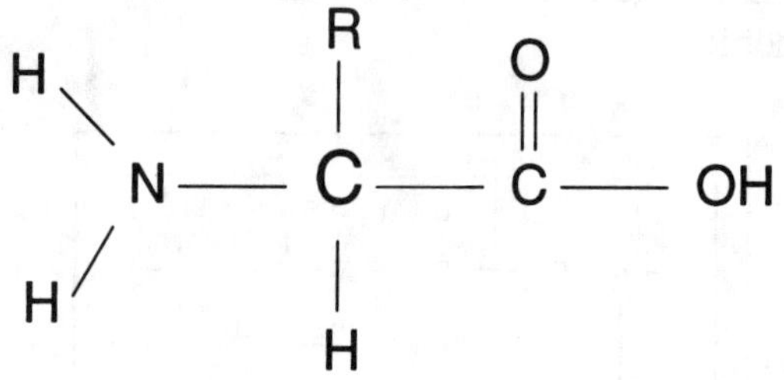

Figure 2.5 Structural Formula of an Amino Acid

Primary Structure - The primary structure is the sequence of amino acids bonded in the polypeptide.

Secondary Structure - The secondary structure is formed by hydrogen bonds between amino acids; the polypeptide can coil into a helix or form a pleated sheet.

Tertiary Structure - The tertiary structure refers to the three-dimensional folding of the helix or pleated sheet.

Quaternary Structure - The quaternary structure refers to the spatial relationship among the polypeptide in the protein.

Proteins have the following biological functions:

Structure - They are found in hair, bones, muscles, and all cell membranes.

Regulation - Some hormones are protein, including insulin. Insulin regulates the level of glucose in the blood.

Transport - Hemoglobin is a protein. It combines with oxygen in red blood cells as they transport oxygen.

Contraction - Actin and myosin are contractile proteins in muscle cells, producing movement.

Catalysts - All enzymes, organic catalysts, are mainly protein.

2.3.4 Nucleic Acids

Nucleic acids are long molecules consisting of bonded subunits called **nucleotides**. A nucleotide consists of three parts: a five-carbon sugar, a nitrogen base, and a phosphate group. The two nucleic acids important biologically are DNA and RNA.

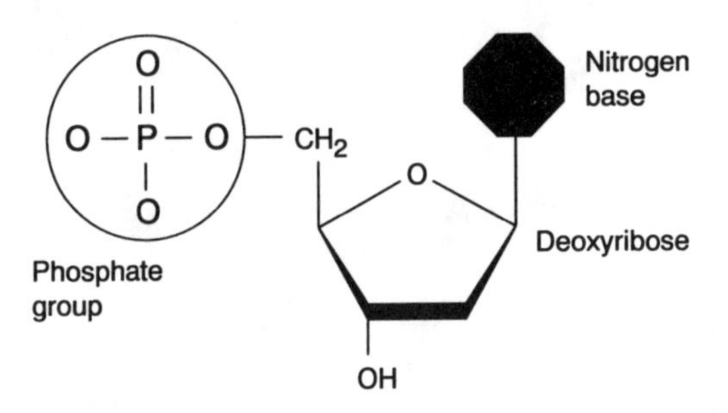

Figure 2.6 Structure Formula of a Nucleotide

DNA - This molecule (deoxyribonucleic acid) consists of two long strands of bonded nucleotides. The base of each nucleotide is either adenine, cytosine, guanine, or thymine. The strands are also linked to each other through hydrogen bonding between base pairs (A-T or G-C). This ladderlike molecule is twisted into the shape of a double helix. DNA is the hereditary material of cells. It determines the development of genetic characteristics.

RNA - This molecule (ribonucleic acid) is a single-stranded molecule of bonded nucleotides. The base of each nucleotide is either adenine, cytosine, guanine, or uracil. DNA directs the synthesis of RNA in the cell. Through this message it determines the synthesis of proteins.

CHAPTER 3

Cells

3.1 Cell Structures

The cell is the smallest unit displaying the characteristics of life. It is the unit of structure and unit of function of the human body. Cells are studied through the light microscope and electron microscope. Observations through the electron microscope reveal details of the membrane, nucleus, cytoplasm, and organelles of cell structure.

3.1.1 Cell Membrane

The cell membrane, or plasma membrane, is the boundary of the cell. This boundary separates the outside of the cell (extracellular environment) from the inside of the cell (intracellular environment). This membrane consists of a double layer of lipid molecules. Proteins are embedded within this lipid bilayer.

The cell membrane regulates the passage of substances into and out of the cell. Therefore, it is a **semipermeable** membrane. Water passes freely through the membrane. Other molecules may pass through depending on their size and the ability of the membrane to recognize and transport them.

3.1.2 Nucleus

The nucleus is defined by a double membrane, the **nuclear envelope**. There are pores through this envelope. The nucleus controls the characteristics and functions of the cell. **Chromatin,** a mass of thread-like material, is found in the nucleus. The chromatin condenses into rodlike structures called **chromosomes** during cell division. Each chromosome carries a series of genes, the units of heredity. The genes are composed of DNA.

The **nucleolus** is an oval body found inside the nucleus. It consists of RNA and protein. Nucleoli are made by the chromosomes and participate in protein synthesis.

3.1.3 Cytoplasm

The cytoplasm is the material found between the nucleus and cell membrane. It consists of water and different organic molecules. Various organelles are embedded in the cytoplasm.

3.1.4 Organelles

The organelles are the specific regions and parts of the cytoplasm. Each organelle has a distinctive structure and function.

Centriole - The centriole is a short cylinder near the nuclear envelope. Normally there are two centrioles at right angles to each other. They coordinate the events of cell division.

Endoplasmic Reticulum - The ER is a membranous series of tubular channels. It is continuous with the nuclear envelope. It provides a passageway for the transport of substances. The rough ER is covered with ribosomes. The smooth ER is not covered with ribosomes.

Ribosome - The ribosomes are small particles consisting of RNA. They are the site of protein synthesis.

Golgi Apparatus - This organelle is a series of flattened vacuoles. They package, store, and modify products that are secreted from the cell.

Mitochondrion - Mitochondria are the powerhouses of the cell. Each one has a double membrane which is the site of chemical reactions that extract energy from nutrient molecules.

Lysosome - The lysosome is a membrane-enclosed organelle containing digestive enzymes. The release of these enzymes can break down substances.

Vacuole - Various membrane-bound vacuoles can store water, nutrients, or waste products in the cell.

Peroxisome - This membrane-bound organelle contains enzymes for oxidation reactions in the cell.

Cytoskeleton - Several kinds of proteins in the cytoplasm form the cytoskeleton. Microfilaments are long thin fibers. Microtubules are thin cylinders. They maintain cell shape and influence cell movement.

Cilium - The cilium is a short, hairlike projection from the cell membrane. The coordinated beating of many cilia produce organized movement.

Flagellum - A flagellum is a long, whiplike organelle extending from the cell membrane. Its action produces movement.

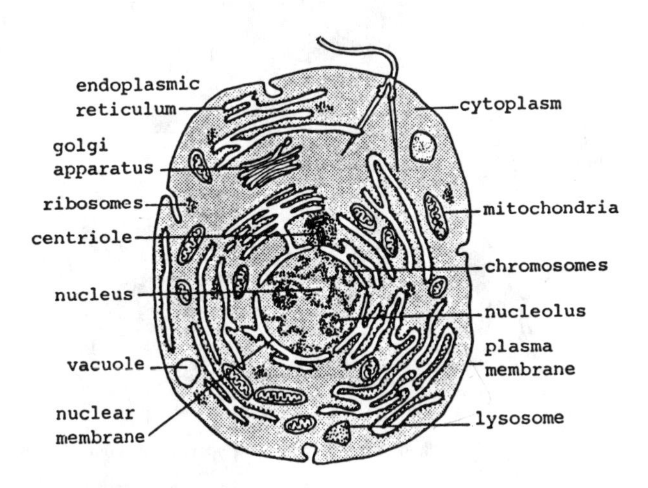

Figure 3.1 Generalized Animal Cell

3.2 Cell Transport

Molecules are transported into and out of the cell. They are also transported between regions within the cell. Transport processes include diffusion, osmosis, active transport, filtration, endocytosis, and exocytosis.

Diffusion - Diffusion is the movement of molecules from a region of higher concentration to a region of lower concentration. Oxygen diffuses from the blood into the cells of the body. Diffusion is a passive process, as it does not require an energy output from the cell.

Osmosis - Osmosis is the diffusion of water through a semipermeable membrane. A cell can lose water to a **hypertonic** medium by osmosis. This medium has a higher concentration of solute molecules. Therefore, its concentration of water is less and water will diffuse toward it. A cell will gain water from a **hypotonic** medium. This medium has a lower concentration of solute molecules. Therefore, its concentration of water is greater and water will leave it to enter the cell.

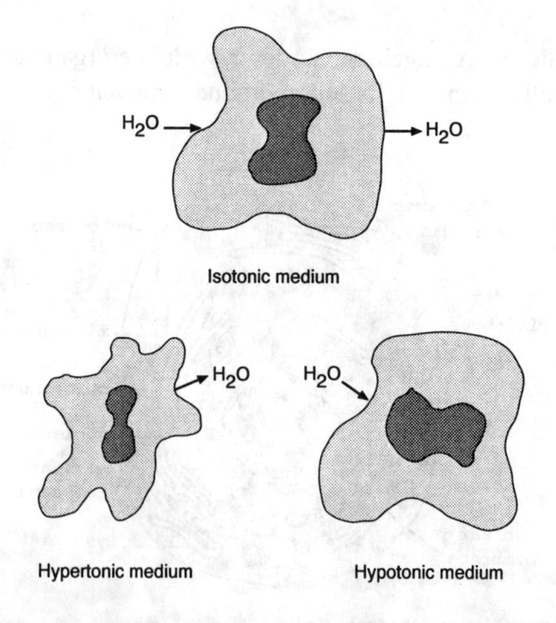

Figure 3.2 Osmotic Effects on a Cell in Hypertonic, Hypotonic, and Isotonic Environments

Normally the extracellular environment is **isotonic** compared to the inside of human cells. This means that the solute concentrations are equal. In this case, the cell does not lose or gain water.

Active Transport - By this process ions and molecules move from a region of lower concentration to a region of higher concentration. The cell must expend energy for active transport. Nerve cells, for example, actively transport sodium ions to the extracellular environment.

Filtration - Molecules pass through membranes by physical force during filtration. For example, blood pressure forces substances across the thin walls of capillaries by filtration. By this process, water and other molecules leave the circulation to serve the cells of the body.

Endocytosis - Through endocytosis the cell actively encloses an extracellular particle, forming a membrane-bound vesicle in the cell. If the particle is solid, this is called **phagocytosis**. If it is a liquid, it is called **pinocytosis**. Leukocytes (white blood cells) protect the body by phagocytosis.

Exocytosis - This is the reverse of endocytosis. The cell discharges membrane-bound substances that were intracellular.

3.3 Cell Reproduction

There are two types of cell reproduction that occur in the human body. Body cells reproduce by mitosis. This type of cell division is necessary for growth and tissue repair. Meiosis is the other kind of cell division. Through meiosis, sex cells are produced.

3.3.1 Mitosis

During mitosis one parent cell divides to produce two daughter cells. Each parent cell in the human body normally has forty-six chromosomes. By mitosis each daughter cell produced will receive the same number and kinds of chromosomes as the original parent cell. The identity of the genes on each chromosome will also be the same.

Mitosis is part of the entire lifespan of the cell, also called the **cell cycle**. This entire cycle consists of the following stages:

Interphase - The nuclear envelope of the cell is intact and the nucleolus is visible. The chromosomes are not visible, appearing as chromatin. Interphase makes up the majority of the cell cycle. The cell is carrying out its normal functions: synthesizing, transporting, and storing substances. It is also preparing for cell division. Duplication of the chromosomes is one of the important steps of preparation.

Prophase - This is the first active stage of mitosis. The nuclear envelope and nucleolus disappear. The chromosomes become distinct, appearing double-stranded. The duplicates of each chromosome are called **chromatids**. The chromatids are attached at a region called the **centromere**. The centrioles migrate to opposite poles of the cell, organizing a mitotic spindle. Near the end of prophase, the chromosomes migrate toward the spindle.

Metaphase - The chromosomes line up and attach along the center of the mitotic spindle. This central region is the **equator**. Each double-stranded chromosome attaches to a different spindle fiber by its centromere.

Anaphase - The centromere of each chromosome splits, separating the duplicates (chromatids) of each chromosome. The duplicates of each chromosome are pulled toward opposite ends of the cell by the contraction of the spindle fibers. **Cytokinesis,** cytoplasmic division, begins.

Telophase - Cytokinesis is complete as the parent cell is pinched into two separate daughter cells. Each daughter cell contains one copy of each chromosome. Therefore the daughter cells are genetically identical. Most of the other events of telophase are the reverse of prophase. For example, the nuclear envelope and nucleolus reappear and the chromosomes disappear.

3.3.2 Meiosis

Mitosis produces **diploid** cells. Diploid means that each kind of chromosome occurs in pairs. In human body cells there are 23 kinds of chromosomes. Therefore, the diploid chromosome number is 46 (23 × 2). By contrast, meiosis produces **haploid** cells. Haploid means that there is only one copy of each chromosome. Therefore, the haploid chromosome number in humans is 23.

Meiosis involves two consecutive divisions from a parent cell, producing four daughter cells. The main events of each meiotic division (meiosis I and meiosis II) are:

Meiosis I - After each chromosome duplicates, the chromosomes of each pair are separated into different daughter cells as the parent cell divides. This reduces the chromosome number from diploid (in the parent cell) to haploid (in each daughter cell). Each of the 23 chromosomes in the daughter cells remains double-stranded.

Meiosis II - Each haploid cell produced from meiosis I divides again. The duplicates (chromatids) of each chromosome are separated and move into different daughter cells. The daughter cells produced by meiosis II are also haploid. However, each of the 23 chromosomes is now single-stranded.

Meiosis in males produces sperm cells. It is called **spermatogenesis**. The meiotic production of female sex cells is called **oogenesis**.

CHAPTER 4

Tissues

4.1 Epithelium

The study of tissues is called **histology.** There are four principal kinds of tissues in the human body: epithelial tissue (epithelium), connective tissue, nervous tissue, and muscle tissue. Each major kind has a variety of subtypes.

Epithelial tissue covers the free surfaces of the body. Externally it protects the body. As a covering for the internal surfaces its functions range from secretion to absorption.

There are three types of epithelium based on the shape of the cells:

Squamous - Squamous cells are flat and thin.

Cuboidal - These cells are cube-shaped.

Columnar - These cells are shaped as columns, with a nucleus at the base of the cell.

Epithelium can be simple or stratified:

Simple - Simple means that the cells exist in one layer as they cover a surface.

Stratified - This means that there are many layers of epithelium covering a surface.

For example, simple squamous epithelium lines the inside of blood vessels. Simple cuboidal epithelium lines the tubules of the kidney where the cells absorb molecules. Simple columnar epithelium lines much of the digestive tract where it absorbs and secretes substances. Stratified squamous epithelium lines the inside of the oral cavity and composes the outer layer of the skin, the epidermis.

Other kinds of epithelium include:

Pseudostratified - It appears to be stratified but the cells form only one layer. This tissue lines much of the respiratory tract.

Transitional - The shape of the cells changes. The epithelial cells lining the inside surface of the bladder change shape depending on the amount of urine stored.

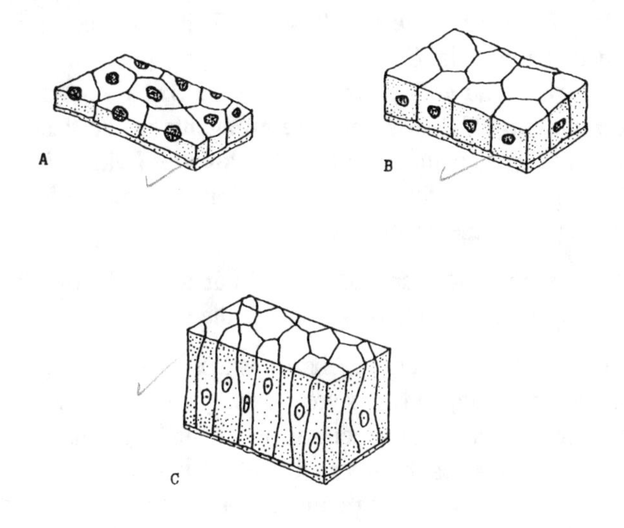

Figure 4.1 Epithelial Cells: Squamous(A), Cuboidal(B), and Columnar(C)

Glandular - Glandular cells secrete. The cells of exocrine glands secrete their products into ducts. Examples are the sweat glands of the skin. Endocrine glands are ductless glands. They secrete chemical messengers, hormones, into the bloodstream.

4.2 Connective Tissue

Connective tissue generally consists of cells which are separated by large amounts of intercellular material that make up the matrix. There is usually an abundant supply of blood vessels throughout the matrix. The matrix may contain several types of fibers:

Collagen - These are white fibers that are flexible and tough.

Elastin - Elastin fibers are yellow and elastic.

Reticular - Reticular fibers are delicate and branching.

Connective tissue binds structures of the body. Other functions include protection, support, and insulation. There are many subtypes of connective tissue.

Areolar (Loose) Connective Tissue - This tissue contains a variety of specialized cells. **Fibroblasts** are large, starshaped cells that produce the white and yellow fibers in the matrix. **Macrophages** are large cells that engulf debris and foreign agents in the tissue. **Mast cells** secrete heparin, an anticoagulant. **Leukocytes** (white blood cells) wander through the matrix to fight infection. **Plasma cells** produce antibodies as part of the immune system.

Areolar tissue is widespread throughout the body. It often lies beneath epithelium. It is abundant beneath the skin.

Dense Fibrous Connective Tissue - Compared to areolar connective tissue, the collagen fibers are closely packed together in dense connective tissue. The arrangement of fibers can be regular or irregular. This tissue composes the internal layer of the skin, the dermis. It is also found in tendons, which connect muscles to bones, and ligaments, which connect bones to bones.

Cartilage - The **chondrocyte** is the cartilage cell. These cells reside in depressions in the matrix, called **lacunae**. A direct blood supply is absent in cartilage. There are three subtypes based on the composition of the matrix:

Hyaline Cartilage - This tissue has fine collagen fibers. It is found in the external nose, in the wall of the trachea, and on the ends of long bones. It reduces friction between the ends of these bones.

Fibrocartilage - This tissue has tough collagen fibers. It is found between the vertebrae of the vertebral column. These wedges of cartilage serve as shock absorbers.

Elastic Cartilage - It has more elastic fibers. It is found in the external ear.

Bone - The matrix of bone tissue is fortified with salts of calcium and phosphorous, making it extremely hard. In compact bone, the **osteocytes** (bone cells) are found in lacunae. The lacunae are arranged in concentric circles around a **Haversian canal**. This canal has blood vessels and nerves serving the bone cells. Miniature canals (canalculi) connect the lacunae with the Haversian canal for transport. This entire system is called a **Haversian system.**

Adipose - This tissue has numerous cells that store fat. It is found beneath the skin where it serves as a layer of insulation. It is also found around the kidneys.

Blood - The **plasma** is the liquid portion of the blood. It is the matrix of this type of connective tissue. **Erythrocytes** (red blood cells) carry oxygen and carbon dioxide in the blood. **Leukocytes** (white blood cells) fight infection. **Thrombocytes** (platelets) are cell fragments that start the process of blood clotting.

4.3 Muscle Tissue

The cells of muscle tissue have the ability to shorten their length, contraction. There are three subtypes of muscle tissue: skeletal, visceral, and cardiac.

Skeletal - This tissue attaches to the bones. As it contracts, it pulls on the bones, producing body movement. The cells, called fibers, are long and threadlike. Each cell has many nucleii (multinucleated) along the inside surface of the cell membrane. Under the microscope this tissue is **striated,** meaning that it has a crossbanded appearance. It is voluntary, capable of rapid response.

Visceral - This muscle tissue is **smooth**, lacking striations, and is involuntary. The cells are small and spindle-shaped, with one nucleus per cell. This tissue is capable of slow, prolonged contractions. It composes the musculature of all internal organs except the heart. Examples include the bladder, uterus, stomach, small intestine, and middle wall of arteries and veins.

Cardiac - This tissue composes the musculature of the heart chambers. It is **striated** and involuntary. There is some branching between cells. **Intercalated disks** separate the cells transversely. Each cell has one nucleus.

4.4 Nerve Tissue

The **neuron** is the specialized cell of nerve tissue. It has the ability to send signals. The signal normally travels from the **dendrites** to the **cell body** to the **axon** of the nerve cell. The cell body contains the nucleus and most of the cytoplasm. The dendrites and axon are processes, or nerve fibers, of the cell. Outside the brain and spinal cord (central nervous system) these processes are found in nerves. Nerves send messages from sense organs to the central nervous system and also send them in the opposite direction to organs that make responses (i.e., skeletal muscles).

Glial cells are another type of cell in nerve tissue. They protect and support neurons.

CHAPTER 5

The Skin

5.1 Functions

The skin is a widespread organ, representing about 3,000 square inches of body surface. It consists of the **epidermis** (outer layer) and **dermis** (inner layer). Its functions include protection, sensory reception, regulation of body temperature, vitamin D synthesis, and identification.

Protection - The skin provides a barrier that protects the entire body. It offers a line of defense against invading bacteria. It serves as a cushion to guard underlying structures from physical forces striking the body. The skin is waterproofed by a thin, oily film secreted by sebaceous glands. Melanin is a dark pigment found in the cells of the epidermis. This pigment screens out ultraviolet rays from the sun that can damage tissues.

Sensory Reception - Receptors are specialized cells that detect environmental changes called stimuli. The layers of the skin contain receptors for touch, pressure, pain, and temperature. Sensory neurons in the skin send signals from receptors to the central nervous system for interpretation.

Regulation of Body Temperature - The dermis contains an abundant supply of blood vessels. The blood carries body heat. If the cutaneous (skin) vessels supplying the skin dilate, more blood reaches the surface of the body through the skin. The heat can escape from these dilated blood vessels. This response serves as an escape from these dilated blood vessels. This response serves as a cooling mechanism. If these supplying vessels constrict, less blood reaches the surface of the body for heat liberation. This response conserves body heat.

The skin also contains sweat glands. As these glands secrete perspiration, heat is required to evaporate this substance from the body surface. This heat is supplied from the body. Therefore, this response cools the body as heat is liberated.

Vitamin D Synthesis - Ultraviolet rays stimulate the synthesis of vitamin D by skin cells. Vitamin D is converted to other substances that regulate the storage of calcium and phosphorous in the bones.

Identification - Papillae are ridges where the epidermis and dermis meet. On the surface these ridges appear as fingerprints and palm prints.

5.2 Structure

The skin is an organ consisting of several tissues. Stratified squamous epithelium composes the outer layer of the skin, the epidermis. Dense connective tissue composes the deeper, thicker layer of the skin, the dermis. A subcutaneous layer is found beneath the dermis. It is composed of areolar connective tissue and adipose tissue. Sometimes it is counted as a third layer of the skin and called the hypodermis.

5.2.1 Epidermis

The epidermis consists of four or five sublayers depending on its location in the body. The layers range from the outermost sublayer, the stratum corneum, to the deepest layer, the stratum basale. Each layer forms cells that become part of the sublayer next to it externally.

The **stratum corneum** consists of squamous epithelial cells that are dead. The cells of this sublayer are cornified, meaning that they are hard and filled with the protein keratin. The cells are dead because they are too far from the nutrients and oxygen supplied from the blood found in the underlying dermis. Therefore, they are constantly being shed from the body surface. Division by cells in the deeper sublayers of the epidermis replace the cells in this layer as they are lost.

The **stratum lucidum** is present if five sublayers are present where the skin is thicker. Examples include the palms of the hand, fingertips, and soles of the feet. When present, this sublayer of dead cells is under the stratum corneum.

The **stratum granulosum** consists of flattened epithelial cells. As these cells are pushed toward the surface, they gradually lose their organelles and die.

The **stratum spinosum** consists of living cells under the stratum granulosum. As they divide, they are pushed toward the granulosum.

The **stratum basale** is the deepest sublayer of the epidermis. It is a single layer of column-shaped cells that are actively dividing. New cells produced are pushed outward where they become part of the stratum spinosum.

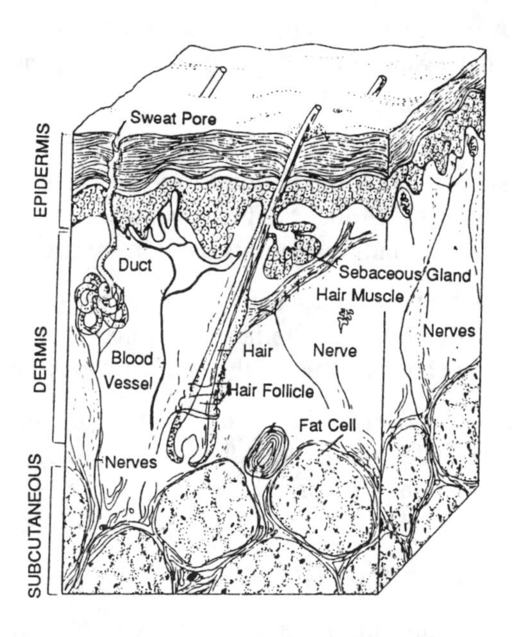

Figure 5.1 Section of the Skin: Epidermis, Dermis, and Hypodermis

5.2.2 Dermis

The dermis contains numerous structures. Many of the functions of the skin are determined by the action of these structures.

The dermis is laced with **collagen** and **elastin** fibers. The collagen provides toughness and the elastin fibers allow the skin to stretch.

Arteries supply blood to the dermis. **Veins** transport blood away from this layer. If the arteries dilate, more blood is supplied to the dermis. This blood carries heat which can escape from the body. If the supplying vessels constrict, body heat is conserved. Therefore, this blood flow contributes to temperature regulation.

The dermis contains numerous **neurons** and **receptors**. By the role of these structures the skin serves as a sensory organ. There are different receptors specialized for different stimuli: touch, pressure, pain, and temperature. Sensory neurons conduct signals from these receptors to the central nervous system.

Sweat glands form and secrete perspiration. The ducts of these exocrine glands continue through the epidermis to the skin surface. The secretion and evaporation of perspiration serves as a cooling mechanism.

Where hair is present in the skin, the root of each hair is anchored in a chamber called the **hair follicle. Sebaceous glands** are associated with these hair follicles. These glands secrete an oily product, sebum, that covers and waterproofs the skin surface.

Small masses of smooth muscle tissue, the **arrector pili** muscles, are also associated with the hair follicles. They contract when external temperatures are cold, producing the familiar "goosebumps."

Papillae are the ridges at the boundary of the dermis and stratum basale of the epidermis. These ridges are pronounced enough to establish fingerprints and palmprints.

5.3 Accessory Structures

Accessory structures originate from the layers of the skin. These include the glands (sweat and sebaceous), hair and nails. The toenails and fingernails consist of a specialized, hardened type of keratin that is produced from epidermal cells.

The skin is often called the integument. The integument and accessory structures are recognized as the **integumentary system,** one of the organ systems of the body.

5.4 Membranes

Membranes are large, sheetlike boundaries that cover the surfaces of the body. There are four kinds of membranes; cutaneous, mucous, serous, and synovial. Each membrane consists of an epithelial tissue mounted on a base of connective tissue.

Cutaneous Membrane - The skin is the cutaneous membrane. Stratified squamous epithelium composes the epidermis. Dense connective tissue composes the dermis.

Mucous Membrane - Mucous membranes line the internal body cavities that are contiguous with the external environment. Examples include the respiratory, digestive, urinary, and reproductive tracts. The epithelial cells of these membranes secrete mucous, which prevents the drying out of the membrane. It also offers lubrication.

Serous Membrane - Serous membranes line body cavities and organs that are sealed off from the external environment.

Serous membranes usually exist at two levels. The **parietal** membrane lines the cavity. The **visceral** membrane lines the organ within the cavity. For example, the parietal pleura is a serous membrane lining the part of the thoracic cavity around the lung. The visceral pleura (pulmonary pleura) adheres to the lung surface. These membranes seal off a cavity around each lung, the pleural (intrapleural) cavity.

Synovial Membrane - Synovial membranes line the cavities of joint capsules at the freely-movable joints. They secrete synovium, an oily product that serves as a lubricant.

CHAPTER 6

The Skeletal System

6.1 Functions

The skeletal system consists of 206 bones that are large enough to be counted. It provides at least five functions for the human body.

Protection - Many organs are contained within spaces formed by the bones. The cranium, which is the superior portion of the skull, houses the brain. The vertebral column surrounds the spinal cord. Twelve pairs of ribs and the sternum, composing the thoracic cage, protect many organs in the ventral body cavity. Examples include the heart, lungs, stomach, and liver.

Support - The vertebral column has four different curvatures along its length. These curvatures give the backbone a great ability to support body weight. The arch formed by the bones of the foot also supports body weight. Other bones, such as the long bones in the legs and arms, contribute great mechanical strength.

Movement - Skeletal muscles attach to the bones. As the muscles contract they pull on the bones and produce movement. For example, the forearm bones (radius and ulna) are pulled toward the upper arm bone (humerus) as the biceps brachii muscle contracts. The forearm bones and humerus are connected by a hinge joint. This joint permits the bending (flexion) of the forearm when the biceps muscle in the upper arm contracts.

Mineral Storage - Calcium and phosphorous are stored in the bones. If the concentration of these minerals is too high in the blood, the excess amount is stored in the bones. If more of these minerals is needed in the blood, they are released from the bones.

Blood Cell Formation (Hemopoiesis) - The bones are not solid structures. Cavities in the cranial bones, vertebrae, ribs, sternum, and ends of long bones contain red marrow. This blood-forming tissue produces erythrocytes (red blood cells), leukocytes (white blood cells), and thrombocytes (platelets). From these sites of production, these cells are released into the circulation.

6.2 Growth and Development

Based on shape, there are five kinds of bones that develop in the body. **Long** bones are found in the arms (i.e., humerus, the upper arm bone) and legs (i.e., femur, the thighbone). **Short** bones include the carpals (wrist) and tarsals (ankle). Some bones are **flat** such as the sternum. Others are **irregular** such as the mandible (jawbone) and vertebrae. **Sesamoid** bones are seed-shaped, found in joints (i.e., patella, the kneecap bone).

Long bones increase in diameter through the activity of cells. Cells called **osteoblasts** on the surface of the bone produce layers of new bone cells, **osteocytes**. These bone cells mature and produce a matrix, surrounding inorganic material, to increase the amount of compact (dense) bone tissue. As this process progresses, the long bone increases in diameter.

The activity of osteoblasts and osteocytes also produces compact bone tissue in the other kinds of bones (short, flat, irregular, and sesamoid).

Long bones also increase in length during growth and development. The **epiphyseal plate** (disc) is a wedge of cartilage accounting for this increase. This plate is found between the epiphysis (bulbous end) and diaphysis (tubular shaft) at each end of the bone. The cartilage cells of the epiphyseal plate form layers of compact bone tissue, adding to the length of the bone. This disc becomes inactive in most individuals by the late teens or early twenties.

In the adult, the skeletal system is constantly being remodeled. Bones are being broken down and rebuilt. **Osteoclasts** are cells that break down and remove worn-out bone tissue. Osteoblasts build new bone tissue to replace this loss.

6.3 Gross Anatomy of a Long Bone

A long bone, such as the femur, can be used to illustrate the gross (large) anatomy of a bone. Its major parts include:

Epiphysis - the bulb-like end proximally and distally

Diaphysis - tube-like shaft between the epiphyses

Metaphysis - line between the epiphysis and diaphysis—this is the earlier site of the epiphyseal plate during bone growth.

Medullary Cavity - a cavity in the diaphysis; this is filled with yellow marrow in the adult. The yellow marrow is mainly fat tissue.

Compact Bone - dense bone tissue composing the wall of the diaphysis

Cancellous Bone - spongy bone tissue in the epiphysis—the spaces of this spongy bone are filled with red marrow that produces blood cells.

Endosteum - the lining of the medullary cavity

Periosteum - the outer covering on the diaphysis—it is necessary for the nutritional maintenance of the long bone.

Articular Cartilage - hyaline cartilage covering the ends of the long bone at the joints (articulations)—this smooth covering reduces friction during movement.

Short, flat, irregular, and sesamoid bones consist of a thin layer of compact bone. Red marrow fills cancellous bone tissue inside the bone.

6.4 Microscopic Anatomy of a Bone

Compact bone tissue is a type of connective tissue. The matrix of this tissue consists mainly of collagen plus salts of calcium and phosphorous. These mineralized salts give the matrix the characteristics of reinforced concrete.

In addition to the matrix, compact bone consists of:

Osteocytes - the bone cells

Lacunae - Each lacuna is a depression in the matrix where an osteocyte is located.

Lamellae - Each lamella is a circular layer of osteocytes located in lacunae.

Canaliculi - processes connecting the lacunae—each canaliculus resembles a miniature canal.

Haversian Canal - This is a central canal around which the concentric lamellae are located. The Haversian canal contains blood vessels and nerves that serve the osteocytes. Exchange of substances (e.g., oxygen, nutrients) between the central canal and osteocytes occurs along the canaliculi connecting the lacunae to the Haversian canal.

The Haversian canal and surrounding structures form a **Haversian system.** This repeating system is found in the compact bone of the diaphysis of a long bone.

Cancellous bone tissue is found in the epiphyses of long bones and inside the short, flat, irregular, and sesamoid bones. It consists of interconnecting plates called **trabeculae**. Each trabecula consists of several lamellae with osteocytes.

6.5 Axial Skeleton

The axial skeleton is one branch of the skeletal system. It forms the midline of the skeleton. This branch consists of the skull, hyoid bone, vertebral column, and thoracic cage. Eighty of the 206 bones of the skeleton are axial.

Skull - Twenty-eight bones are found in the skull. Eight bones make up the **cranium** or superior portion of the skull. They are the **frontal**, **temporal** (2), **parietal** (2), **occipital**, **ethmoid**, and **sphenoid** bones. The cranium houses the brain.

The 14 bones of the face are the **maxillae** (2), **zygomatic** (2), **nasal** (2), **lacrimal** (2), **palatine** (2), **inferior nasal conchae** (2), **vomer**, and **mandible** (jawbone). The mandible is the only movable bone of the skull. All others are connected by immovable joints called sutures. Three middle-ear bones are found in each temporal bone (See Chapter 9).

Hyoid Bone - This is a horseshoe-shaped bone suspended by muscles from the floor of the oral cavity.

Vertebral Column - The backbone consists of four curvatures: **cervical** (neck), **thoracic** (chest), **lumbar** (lower back), and **pelvic**. Each consists of a serial arrangement of vertebrae. These vertebrae form a continuous tube, housing the spinal cord.

The cervical curvature has seven vertebrae. The **atlas** is the first cervical vertebra (C1), supporting the skull. The **axis** is the second cervical vertebra (C2). The seventh cervical vertebra is the most inferior.

Twelve thoracic vertebrae compose the thoracic curvature. The first thoracic vertebra is most superior. The twelfth vertebra is the most inferior one.

Five lumbar vertebrae make up the lumbar curvature. From the base of the brain, the spinal cord inside the backbone ends inferiorly at the first lumbar vertebra. The **sacrum** (five fused vertebrae) and **coccyx** (tailbone, four fused vertebrae) make up the pelvic curvature. The sacrum is directly under the fifth lumbar vertebra. The coccyx is the most inferior part of the vertebral column.

Discs of fibrocartilage (**intervertebral discs**) are found between the bodies (anterior portions) of the vertebrae. They act as shock absorbers, adding springiness to the backbone. **Intervertebral foramina** between the vertebrae are openings for the passage of blood vessels and spinal nerves communicating with the spinal cord.

Thoracic Cage - The thoracic cage consists of the **sternum** and twelve pairs of **ribs**. The sternum (breastbone) consists of three pieces: **manubrium** (most superior), **body**, and **xiphoid process** (most inferior).

Each rib pair (1-12) articulates posteriorly with the thoracic vertebra (12) of the same number. Anteriorly, the **vertebrosternal** ribs

(1-7) articulate with the sternum. The **vertebrochondral** ribs (8-10) articulate with the cartilage connecting rib pair seven with the sternum. The **vertebral** ribs (11-12) articulate only with the vertebrae.

Rib pairs 1 through 7 are also called the **true** ribs. Rib pairs 8-12 are called the **false** ribs.

6.6 Appendicular Skeleton

The appendicular skeleton is the other branch of the skeletal system. It contains 126 bones. This branch consists of the **pectoral** (chest) **girdle** and arm bones as well as the **pelvic** (hip) **girdle** and leg bones.

Pectoral Girdle - Each half of the pectoral girdle consists of the **clavicle** (collarbone) and **scapula** (shoulder blade). The medial end of the clavicle articulates to the manubrium of the sternum. Laterally it attaches to the acromion of the scapula.

Arm Bones - The **humerus** is the upper arm bone. The head at the proximal end of each humerus articulates with the glenoid fossa, a shallow depression, of the scapula.

In anatomical position, the **radius** is the lateral bone of each forearm. The **ulna** is the medial forearm bone.

Each wrist consists of eight **carpal** bones, a proximal and distal row of four each.

The **metacarpals** are five bones in the palm of each hand.

The **phalanges** (sing., phalanx) are the bones of the fingers. Two are in the thumb. Three are in each of the other fingers.

Pelvic Girdle - The pelvic girdle consists of the two **os coxa** or hip bones. They connect anteriorly by a slightly movable joint, the **symphysis pubis**. Combined posteriorly with the sacrum and coccyx of the axial skeleton, they form the **pelvis**.

Leg Bones - The **femur** is the thigh bone. The head at the proximal end of each femur fits into the acetabulum, a shallow depression of the hip bone. Distally the femur joins the tibia (shinbone).

The **patella** is the kneecap bone. Each is connected distally to the tibia by a ligament. Proximally it attaches to the quadriceps muscle on the front of the thigh by a tendon.

The **tibia** is the thicker, medial bone of each calf. The **fibula** is the lateral, thinner bone.

Seven **tarsal** bones compose each ankle and the posterior portion of the foot.

Five **metatarsals** compose the arch of each foot.

The **phalanges** are the bones of the toes. Two are in the big toe. Three are found in each of the other toes.

6.7 Articulations

Articulations (joints) are the structures where bones connect. There are three main classes based on the amount of motion they allow:

Synarthroses - A synarthrosis is an immovable joint (fibrous joint). One example is a suture. The parietal bones are locked together by a **sagittal suture**. The frontal bone is united to each of the parietal bones by a **coronal suture**.

Amphiarthroses - An amphiarthrosis is a joint permitting slight mobility. It is also known as a cartilaginous joint. One example is the symphysis pubis, where the two os coxa bones join anteriorly.

Diarthroses - A diarthrosis allows free mobility. It is also called a synovial joint. There are six subclasses within this main class:

ball and socket - The head of one bone (i.e., femur) fits into a shallow depression (i.e., glenoid fossa of the scapula). Motion occurs here in three different planes.

hinge - One example is the knee joint. Another is the elbow joint.

pivot - The ring of one bone rotates around the process of another. The atlas pivots on the axis.

gliding - Bones can slide over each other. Examples occur among the carpal and tarsal bones.

saddle - The bones have a saddle shape. Examples occur between the carpals and metacarpals.

condyloid - An oval-shaped condyle of one bone fits into an elliptical shape of another. The metacarpals join the phalanges.

CHAPTER 7

The Skeletal Muscles

7.1 Functions

Skeletal muscles carry out several important body functions.

Movement - Skeletal muscles attach to the bones. As a muscle contracts it pulls on a bone (or bones) to produce movement. For example, the hamstring muscles are found in the posterior thigh. As each contracts it pulls the calf toward the thigh, producing flexion (bending) of the calf. As it contracts, a muscle pulls an **insertion bone** (movable end) toward an **origin bone** (fixed end). In this example the tibia represents the insertion and the femur represents the origin.

Heat Production - Numerous chemical reactions in muscle cells liberate heat. This heat contributes to the maintenance of body temperature.

Posture - Some skeletal muscles pull on the vertebral column and other parts of the skeleton, helping to maintain an upright stance and posture of the body.

7.2 Structure of a Skeletal Muscle

There are over 600 skeletal muscles in the human body. Each one has the same basic structure at several levels of organization: organ (the muscle), tissue (striated, skeletal), cell (fiber), organelles (e.g., mitochondria and myofibrils), and molecules (e.g., water, actin, myosin).

As an organ a skeletal muscle consists of several tissue types. Skeletal muscle **fibers** are long, threadlike cells that compose skeletal (striated) tissue. These cells have the ability to shorten their length or contract.

Dense fibrous connective tissue (fascia) weaves through a skeletal muscle at several different levels. The **epimysium** is the connective tissue layer that envelopes the entire skeletal muscle. The **perimysium** is a continuation of this outer fascia, dividing the interior of the muscle into bundles of muscle cells. The bundle of cells surrounded by each perimysium is called a fasciculus. The **endomysium** is a connective tissue layer surrounding each muscle fiber.

The interior of a muscle also contains the axons of motor neurons. Each axon signals a group of muscle fibers. This association (one neuron, group of muscle fibers) is called a **motor unit**.

The muscle also contains an abundant supply of blood vessels. The blood delivers glucose and other nutrients, along with oxygen, they are needed for cell metabolism. There are also fat deposits that store energy.

Muscle cells contain many organelles common to most cells (i.e., mitochondria). **Myofibrils** are organelles that establish the contractile ability of muscle cells. Each myofibril is a linear succession of boxlike units called **sarcomeres**. Each sarcomere contains the contractile proteins (myofilaments), **actin** and **myosin**. The myosin is the thicker central protein. The actin is the thinner protein at each end of the sarcomere.

Actin and myosin are organized in each sarcomere in the following regions:

A Band - actin and myosin

H Zone - myosin only, within the A band

I Band - actin only

Z Line - boundary of the sarcomere where actin molecules are attached at each end

7.3 Mechanism of Muscle Contraction

A skeletal muscle contracts by the following series of steps:

1. A motor nerve signals the muscle. Some of the neurons in the nerve develop electrical impulses which signal some fibers in the muscle. Each axon secretes a **neurotransmitter** (chemical signal) called **acetylcholine** at the synapse (**motor end plate**) between the neuron and some muscle fibers. This signal excites each muscle cell.

2. An electrical signal spreads out along the **sarcolemma** (cell membrane) of each muscle cell that is signaled.

3. This signal continues transversely into the sarcoplasm (cytoplasm) of each muscle cell along the membranes of the **T** (transverse) **tubules**.

4. The T tubules join the **SR (sarcoplasmic reticulum)** in the **sarcoplasm** (cytoplasm). The signal spreads from the T tubules to this tubular SR, releasing **Ca ions**.

5. The release of Ca from the SR blocks the action of **troponin**, a protein in the myofibrils. Troponin normally inhibits the interaction of actin and myosin, contractile proteins in the sarcomeres of the myofibrils.

6. With troponin inhibited, actin and myosin can interact. Crossbridges on the myosin slide the actin molecules toward the center of the sarcomere. As the actin molecules are attached to the Z lines, boundaries of the sarcomeres, this shortens the sarcomeres according to the **sliding filament theory**. The actin molecules slide between the central myosin molecules.

7. **Hydrolysis** of **ATP** in the cells, into ADP and phosphate, releases energy to drive the sliding of the filaments (actin and myosin). ATP is rebuilt from an energy-storage compound, creatine phosphate.

8. If enough sarcomeres shorten, myofibrils shorten. If enough myofibrils shorten, the fibers (cells) shorten. If enough fibers shorten, the muscle shortens or contracts.

Each muscle responds by an **all-or-none** law. If stimulated sufficiently, it contracts fully. The force of contraction from an entire muscle depends on the percentage of cells that are active, each cell responding by all or none.

7.4 Patterns of Muscle Contraction

There are several patterns of contraction of a skeletal muscle.

Tonus - Some motor units in a muscle are usually active when a muscle is not contracting enough to produce movement. The muscle remains taut by this tonic response. Tonus (muscle tone) of some muscles maintains posture.

Isometric/Isotonic - By isometric response, a muscle does not contract enough to produce motion (isometric = equal length). It stabilizes a body part. For example, muscles in the shoulder act isometrically if the arms push against an immovable wall. By isotonic response, a muscle does shorten and produces motion.

Isotonic responses can be graphed as several patterns:

Simple Twitch - quick jerky contraction to a single stimulus

Summation - the addition of simple twitches from repeated stimulation of a muscle—the twitches add together (summate) toward a more powerful, unified response. This often occurs when the nervous system signals a muscle at a faster rate, causing the twitches to merge.

Tetanus - the powerful, sustained contraction of a muscle from the summation of simple twitches. Isotonic responses of skeletal muscles in the body are usually tetanic.

Fatigue - The muscle cannot respond when stimulated, as it is exhausted of nutrients and accumulates waste products. The muscle can be depleted of glucose, a source of energy. Lactic acid is a waste product that builds up in muscle cells during anaerobic conditions, which means an absence of oxygen. This occurs when an overworked muscle does not receive oxygen rapidly enough to meet its metabolic needs.

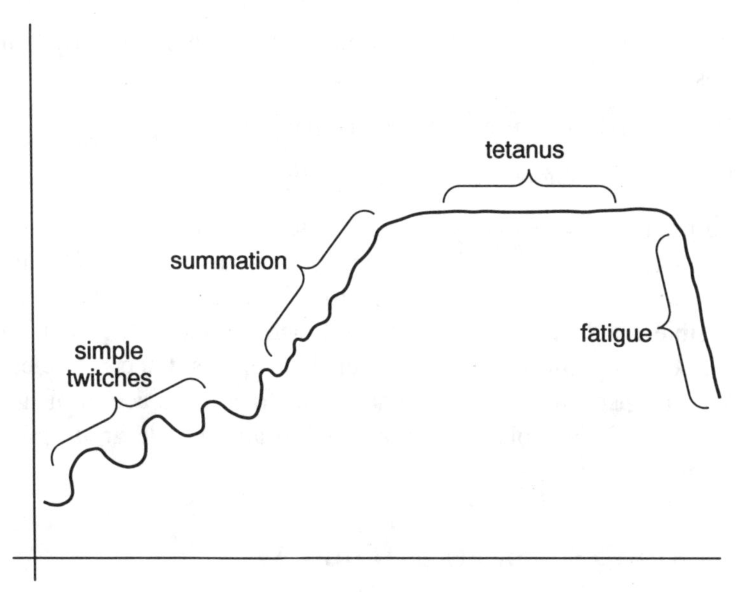

Figure 7.1 Graph of Simple Twitches, Summation, Tetanus, and Fatigue

7.5 Motions

Skeletal muscles attach to bones by tendons. At movable joints (diarthroses), a contracting muscle can pull on a bone (or bones) to produce one of the following motions (actions):

Flexion - bending, decreasing the angle at a joint

Extension - straightening out, increasing the angle at a joint

Abduction - moving a limb away from the midline of the body

Adduction - moving a limb toward the midline of the body

Rotation - pivoting a structure—the atlas (C1) and skull can rotate on the axis (C2). A special case of rotation is pronation and supination. Through **pronation**, the palm is turned posteriorly (radius crosses over the ulna). Through **supination**, it is turned to an anterior position (radius and ulna become parallel).

Dorsiflexion - The toes are lifted as the body is supported on the heel.

Plantar Flexion - The heel is lifted and the body is supported on the toes.

Inversion - The toes are pointed medially.

Eversion - The toes are pointed laterally.

To produce a given body motion, muscles work in groups. Each member has a specific role. For forearm flexion, as one example, the biceps brachii is the **prime mover** by role. It contracts and is mainly responsible for the action. The triceps brachii is the **antagonist** by role. It could oppose the prime mover, but relaxes. Other muscles, called **synergists** by their role, contract to support the action of the prime mover. These roles can be applied to any muscle group producing an action.

7.6 Naming of Skeletal Muscles

The skeletal muscles are named according to the following characteristics. Often two or more of these are applied to the name of the muscle.

Location - The tibialis anterior is in front of the shinbone.

Number of Attachments - The biceps brachii has two origins (fixed ends of attachments). The triceps brachii has three origins.

Direction of Fibers - The fibers of the rectus abdominus muscles range straight (rectus = straight) or parallel to the long axis of the body. The fibers of the external oblique range at an angle (oblique) to this axis.

Shape/Size - The deltoid has the shape of a triangle (delta, triangle). The gluteus maximus of the buttocks is larger compared to the gluteus mininus.

Action - The extensor muscles in the forearm straighten out the fingers. The flexor muscles in the forearm bend the fingers.

A muscle name can reveal two characteristics. The biceps brachii and triceps brachii are found in the upper arm (brachion = arm). The rectus abdominis composes part of the abdominal wall. The external oblique is superficial compared to the internal oblique of the abdominal wall.

7.7 Skeletal Muscles - Body Regions

Some major superficial (surface) skeletal muscles throughout the regions of the body are:

Buccinator - in the cheek—it contracts to compress the cheeks.

Masseter - inserting on the posterior region on the mandible—it raises the mandible, contributing to mastication (chewing).

Temporalis - fan-shaped muscle covering the temporal bone—it supplements (synergistic) the action of the masseter.

Orbicularis Oculi - circular muscle around the anterior margin of the orbit (eye socket)—it closes the eye.

Orbicularis Oris - circular muscle around the anterior margin of the mouth—it contracts to pucker the lips.

Sternocleidomastoid - diagonal muscle on each side of the neck, ranging from sternum and clavicle (origins) to the mastoid process of the temporal bone (insertion)—when both contract, the head is flexed toward the chest.

Deltoid - bulging muscle that covers each shoulder—it abducts the upper arm.

Bicep Brachii - major muscle of the anterior upper arm—it flexes the forearm.

Triceps Brachii - major muscle of the posterior upper arm—it extends the forearm.

Pectorlais Major - major chest muscle—it adducts the upper arm with some medial rotation.

Rectus Abdominis - midline, segmented muscle of the abdominal wall—it flexes the trunk.

External Oblique - located on each side of the rectus abdominis—each one bends the trunk to that side of the body.

Trapezius - large, diamond-shaped muscle of the upper back—it extends the head and works to shrug the shoulders.

Latissimus Dorsi - wide muscle of the lower back—it extends and adducts the upper arm. Rotates humerus posteriorly, shoulder down + posteriorly

Gluteus Maximus - the largest muscle of the buttocks—it extends the thigh. O= Sacrum, coccyx + post. surface of ilium
I - post surface of femur + fascia of thigh

Rectus Femoris - located along the front of the femur—it flexes the thigh. It combines with three other thigh muscles (**vastus lateralis**, **vastus intermedius, vastus medialis**) to form the **quadriceps** muscle. They act to extend the calf.

Sartorius - long, ribbonlike muscle ranging over the front of the thigh—it is used to cross the legs.

2 origins
Biceps Femoris and Semitendinosus - the lateral and medial hamstring muscles respectively, located on the posterior thigh—they flex the calf, extends thigh

Tibialis Anterior - along the front of the tibia—it produces dorsiflexion.

Gastrocnemius - the bulging posterior calf muscle—it produces plantar flexion.

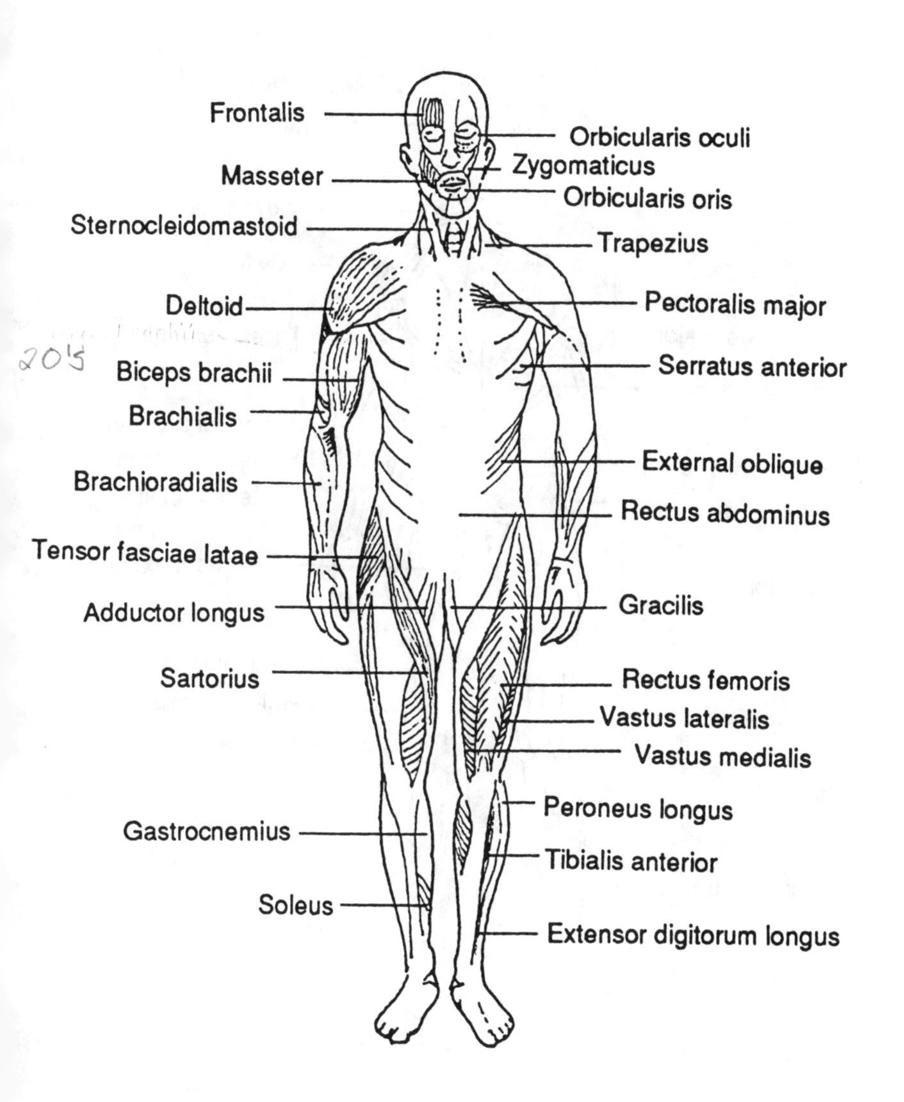

Figure 7.2.A Superficial Muscles—anterior

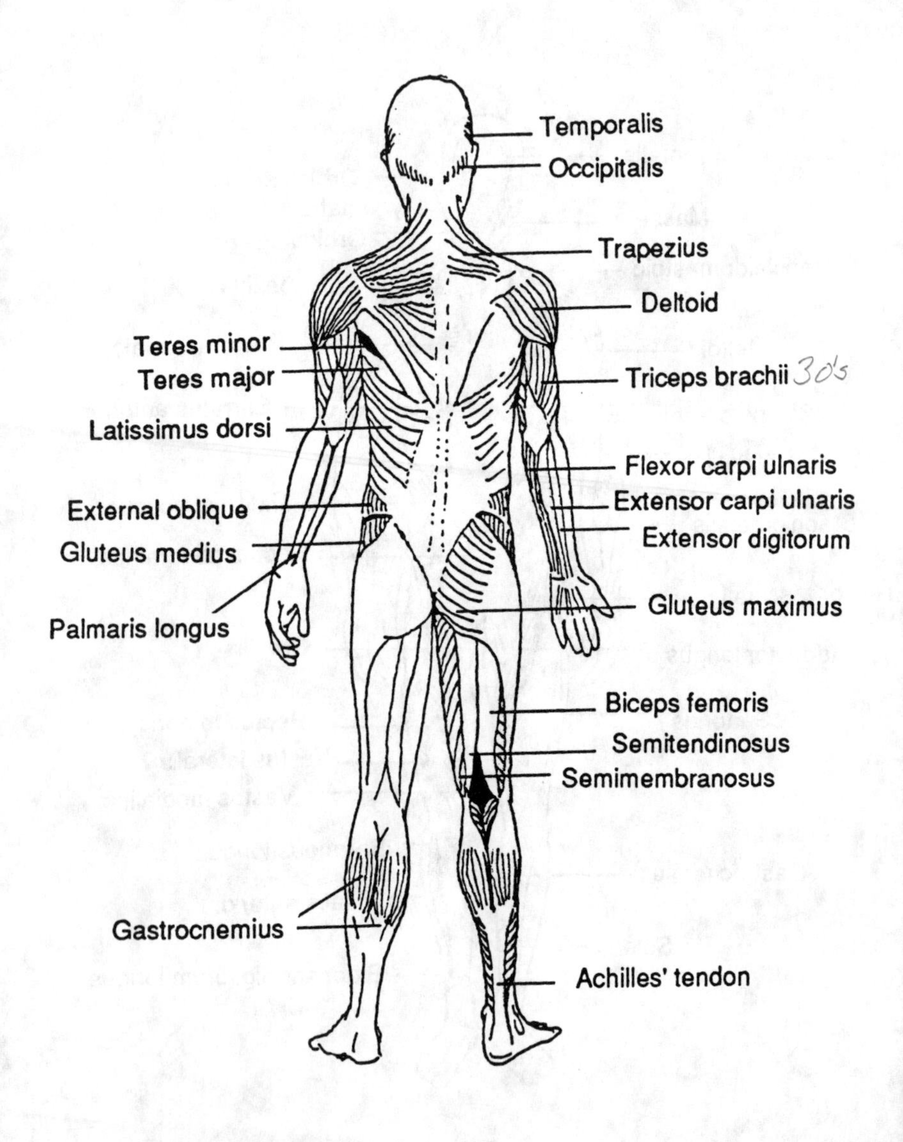

Figure 7.2.B Superficial Muscles—posterior

CHAPTER 8

The Nervous System

8.1 Divisions of the Nervous System

The nervous system is a major communication network that sends signals throughout the body. It consists of two major divisions. The two divisions are connected and work together.

Central Nervous System (CNS) - This division consists of the **brain** and **spinal cord**. The brain is contained within the cranium. The spinal cord is found within the vertebral column, extending from the base of the skull to the first lumbar vertebra.

Peripheral Nervous System (PNS) - This division consists of 12 pairs of **cranial nerves** and 31 pairs of **spinal nerves**. They are continuous with the brain and spinal cord, respectively. However, as branches are connected to the trunk of a tree, they are outside the central branch.

The **autonomic nervous system** is a subdivision of the PNS. It controls motor functions of the internal organs (viscera). It has two branches, the sympathetic and parasympathetic. These two branches have opposing effects on the activity of an organ. All remaining peripheral control is carried out by the **somatic nervous system**, the other branch of the PNS.

8.2 Neuron/Glial Cell

The **neuron** is the cell of the nervous system that sends impulses. **Glial cells** protect and provide support for the neurons. They also provide nourishment.

The neuron has the following regions and specializations:

Cell Body - contains the nucleus and most of the cytoplasm

Dendrite - This process sends the impulse toward the cell body. There may be one or many dendrites per cell. Some neurons lack dendrites.

Axon - This process sends the impulse away from the cell body. There is only one axon per neuron. It is also called the nerve fiber.

Myelin Sheath - This is a white, fatty covering around the axon. It is produced in the PNS by a type of glial cell, the **Schwann cell**, that wraps around the axon. It deposits the myelin around the axon in a series of circular layers. Masses of axons that are myelinated in the nervous system compose the white matter. Unmyelinated axons, plus dendrites and cell bodies, compose the gray matter.

Nodes of Ranvier - There are gaps between Schwann cells along the axon. Where these cells are absent, there are also gaps in the myelin covering the axon, the nodes.

Neurilemma - This is a thin covering along the axon that is external to the myelin when present. It promotes the regeneration of the axon.

There are several kinds of neurons.

Sensory (Afferent) - This neuron is found in nerves and sends signals toward the CNS. It has long dendrites and a short axon.

Motor (Efferent) - This neuron is found in nerves and sends signals away from the CNS. It has short dendrites or the dendrites may be absent. The axon is long.

Interneuron - This cell is found between sensory and motor neurons. It is contained within the CNS.

The several kinds of neurons work together to establish circuits throughout the body. They are separated by junctions between them called **synapses**. For example, a painful stimulus is conducted to the CNS by sensory neurons. A response to this, such as pulling the arm away from the stimulus, is controlled by motor neurons. Interneurons between the sensory and motor neurons complete the circuit needed for this action.

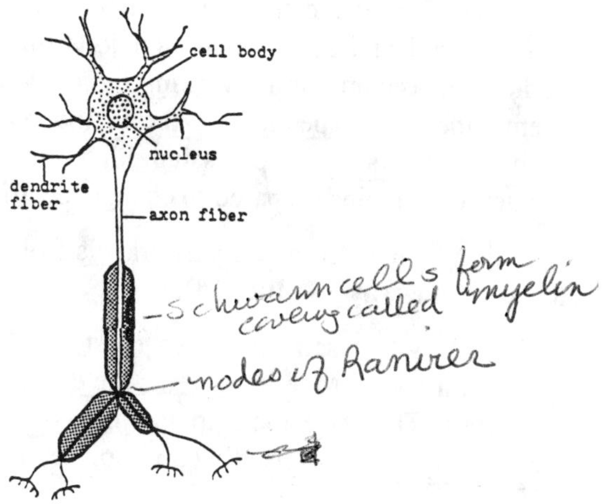

Figure 8.1 The Neuron

8.2.1 Nerve Impulse

The nerve impulse, called the **action potential**, is electrical in nature. It develops by the following events.

1. The **resting membrane potential** occurs in a neuron when it is not firing an impulse. The cell is positive extracellularly and negative intracellularly. It is positive on the outside due to a higher concentration of sodium ions.

2. When stimulated the cell membrane becomes more permeable to sodium ions. These ions are positive. Some of these ions diffuse into the cell. This change travels point by point along the cell membrane, reversing the polarity of the cell (wave of depolarization). The electrical change (positive on the inside, negative on the outside) is called the action potential.

3. After the impulse has passed along the neuron, the normal polarity of the cell is restored. The membrane becomes more permeable to potassium ions, which are higher in concentration intracellularly. They diffuse to the outside, point by point, along the cell membrane. This reestablishes the resting membrane potential.

Each time the neuron fires an impulse, some sodium diffuses into the cell and some potassium ions diffuse to the outside. The sodium-potassium pump, working by active transport within the cell membrane, is independent of the action potential. It keeps sodium high on the outside and potassium high on the inside. By this activity the neuron remains prepared to change electrically to fire impulses.

Along a neuron the impulse travels from dendrite(s) to cell body to axon (**one-way conduction**).

Neurons that send impulses most rapidly are large and myelinated with nodes. Myelin prevents the diffusion of sodium and potassium ions. Therefore the impulse jumps from node to node, where the myelin is absent, speeding up impulse transmission.

8.2.2 Synaptic Transmission

The nature of the signal at the synapse is chemical. It develops by the following steps.

1. A neuron (presynaptic) develops an action potential.
2. The action potential triggers the release of **neurotransmitter** molecules (chemical signal) from the axonal end of this neuron. The neurotransmitter, such as acetylcholine (ACh), is stored in vesicles in the axon.
3. The neurotransmitter diffuses across the synapse from the presynaptic cell.
4. The neurotransmitter excites the next cell, located on the other side of the synapse (postsynaptic).
5. Sodium begins to diffuse into the postsynaptic cell, starting the development of its action potential.

6. The neurotransmitter is broken down or transported away from the synapse. The release of neurotransmitter again is necessary for another signal across the synapse.

8.3 Reflex Arc

The reflex arc is a simple neural pathway connecting receptors to an effector. A receptor detects a stimulus or environmental change. An effector is an organ of response (i.e., skeletal muscle). It produces a response called a reflex. This response does not involve the conscious involvement from higher centers in the brain. The reflex arc has five components that are activated in the following order:

Receptor - This specialized cell, or group of cells, detects a stimulus. During the knee-jerk (patellar) reflex, the patellar ligament (linking the patella to the tibia) is struck. This excites receptors in the tendon connecting the patella to the quadriceps muscle in the anterior thigh.

Sensory Neuron - This neuron sends impulses from the point of the stimulus to the CNS. For the knee-jerk reflex, and other spinal cord reflexes, sensory neurons enter the dorsal region of the cord.

Interneuron - This neuron in the CNS sends impulses from the sensory neuron to the motor neuron. The interneuron is separated from the other two neurons by synapses. For spinal cord reflexes, interneurons are located in the inner zone of gray matter.

Motor Neuron - This neuron sends impulses to the effector. For spinal cord reflexes, motor neurons exit from the ventral region of the spinal cord.

Effector - This organ carries out a response. During the knee-jerk, the quadriceps muscle extends the lower leg.

Reflexes - Automatic responses that depend on the activation of the components of the reflex arc. Some are coordinated by the spinal cord (i.e., knee jerk); others are coordinated by unconscious centers of the brain (i.e., dilation and constriction of the pupil of the eye).

8.4 Central Nervous System

The central nervous system consists of the brain and spinal cord. The brain consists of several regions: hindbrain (rhombencephalon), midbrain (mesencephalon), and forebrain (prosencephalon).

8.4.1 Brain

The **hindbrain** consists of the medulla oblongata, pons, cerebellum, and fourth ventricle.

Medulla Oblongata - Continuous with the spinal cord, this is the most inferior portion of the brain. It contains vital centers - the respiratory, cardiac, and vasomotor (blood pressure) centers. It also coordinates many reflexes - vomiting, swallowing, coughing, and sneezing.

Tracts - This is a group of axons in the CNS that pass through the spinal cord and brain regions. The tracts, ascending and descending, are responsible for vertical relay throughout the CNS. Most of these tracts cross over at the level of the medulla.

Pons - This is a bulbous region that is superior to the medulla. It contains the tracts for vertical relay.

Cerebellum - The cerebellum is posterior to the medulla and pons. This butterfly-shaped structure is responsible for motor coordination. It also controls body equilibrium, posture, and muscle tone.

Fourth Ventricle - This is one of four such chambers throughout the brain. Each secretes and circulates cerebrospinal fluid (CSF).

The **midbrain** consists of the cerebral peduncles (two), the cerebral aqueduct, and the corpora quadragemina.

Cerebral Peduncle - This is the stalklike, anterior portion of the midbrain. Each peduncle contains tracts for vertical relay.

Corpora Quadragemina - This consists of four rounded structures called colliculi (sing., colliculus) that coordiate visual and auditory reflexes.

Cerebral Aqueduct - This is a passageway conducting the CSF from the third ventricle (in the forebrain) to the fourth ventricle.

The **brainstem**, a vertical and unconscious portion of the brain, consists of the medulla, pons, and midbrain. The **forebrain** is mounted on it, much as a cap of a mushroom.

The **lower forebrain** (diencephalon) consists of the thalamus, hypothalamus, and third ventricle.

Third Ventricle - This chamber receives the CSF from the lateral ventricles in the higher forebrain (cerebrum).

Thalamus - The thalamus is a mass of gray matter ranging through the roof of the third ventricle. It contains tracts for vertical relay. Serving as a screen or filter, it allows only some ascending signals to reach the higher forebrain.

Hypothalamus - This is a mass of gray matter forming the lateral walls and floor of the third ventricle. It contains centers for thirst, hunger, body temperature, and water balance. It signals the pituitary gland which is attached to the floor of the third ventricle.

The **cerebrum** is the **higher forebrain**. It consists of two halves or cerebral **hemispheres.** Its main features are:

Cortex - The cerebral cortex is the thin, wrinkled gray matter covering each hemisphere. The cortex has **convolutions** or gyruses, which are elevated areas of the cortex. A **sulcus** is a shallow groove between convolutions. A **fissure** is a deep groove. These markings greatly increase the surface area of the cortex.

Lobes - The cortex of each hemisphere is divided into four lobes. Each lobe has convolutions with mapped **sensory** and **motor** functions: **frontal** (motor - skeletal muscles, speech), **parietal** (sensory - skin and skeletal muscles), **temporal** (sensory - olfaction, taste, hearing), and **occipital** (sensory - vision). **Association areas** make up most of the cortex. They integrate functions between the sensory and motor areas, contributing abstract functions such as memory and reasoning.

White Matter - Each cerebral hemisphere contains a thicker core of white matter, containing myelinated axons that send signals. The

corpus callosum is a bridge of white matter connecting the two hemispheres. Masses of gray matter (i.e., caudate nucleus) are located within the white matter. They control unconscious motor activity.

Ventricles - Each hemisphere contains a **lateral ventricle**. Each chamber is inferior to the corpus callosum and secretes CSF through a foramen into the third ventricle.

8.4.2 Spinal Cord

The spinal cord consists of two halves, divided by an anterior fissure and posterior sulcus. It contains regions of white matter and gray matter. Each performs one function.

White Matter - The white matter is external. On each side of the cord, these myelinated axons are organized into **columns**: **anterior**, **lateral**, and **posterior**. The columns contain tracts for vertical relay. For example, the spinothalmic is an ascending tract. The corticospinal is a descending tract.

Gray Matter - The gray matter is internal, appearing much as a butterfly in flight. On each side it is organized into **horns**: **anterior**, **lateral**, and **posterior**. The horns contain interneurons that complete reflex arcs coordinated by the cord.

The brain and spinal cord are covered by several layers of **meninges** (sing., **meninx**): the **dura mater** (external layer), **arachnoid mater** (middle layer), and **pia mater** (internal layer adhering to nerve tissue).

8.5 Peripheral Nervous System

Among the 12 pairs of cranial nerves, some contain only motor neurons. Some are purely sensory. Others are mixed, containing sensory and motor neurons.

8.5.1 Cranial Nerves/Spinal Nerves

The cranial nerve pairs are numbered one through twelve in a posterior direction as they connect to the brain. Their main functions are:

I. **Olfactory** - sense of smell, sensory nerve

II. **Optic** - sense of vision, sensory nerve

III. **Oculomotor** - moves the eye, motor nerve

IV. **Trochlear** - moves the eye, motor nerve

V. **Trigeminal** - sensory for teeth, eyes, tongue; motor for muscles of mastication (chewing)

VI. **Abducens** - moves the eye, motor nerve

VII. **Facial** - sensory for taste; muscles for facial muscles and salivary glands

VIII. **Vestibulocochlear** - balance and hearing, sensory nerve

IX. **Glossopharyngea** - sensory for tongue; motor for tongue

X. **Vagus** - sensory and motor for internal organs

XI. **Spinal Accessory** - motor of muscles of neck and shoulder

XII. **Hypoglossal** - motor for the tongue

Spinal Nerves - All 31 pairs of spinal nerves are mixed. They are organized into the following groups:

Cervical - 8 pairs

Thoracic - 12 pairs

Lumbar - 5 pairs

Sacral - 5 pairs

Coccygeal - 1 pair

Branches of many of the spinal nerves form complex networks called **plexuses**. For example, the cervical plexus (pairs C1 through C4) controls the skin and muscles of the neck plus the diaphragm. The brachial plexus (C5 through T1) controls the skin and muscles of the arms.

Spinal nerve-spinal cord functions on one side of the body are controlled by the cerebral cortex on the opposite side. For example, the right hemisphere controls the left arm. This is because the tracts for vertical relay through the CNS cross over at the medulla.

8.5.2 Autonomic Nervous System

The autonomic nervous system (ANS) controls motor functions of the internal organs (smooth, cardiac muscle) and glands. It consists of two branches:

Sympathetic - This branch includes fibers of thoracic and lumbar nerves. The nerves have preganglionic and postganglionic neurons. The postganglionic neurons discharge a neurotransmitter called **norepinephrine** at the synapses meeting internal organs.

Parasympathetic - This branch includes fibers of cranial and sacral nerves. The nerves also have preganglionic and postganglionic neurons, but the postganglionic cells discharge **acetylcholine** as the terminal transmitter.

Both branches control the activity of an internal organ at any time. However, the sympathetic branch dominates during the **"fight or flight"** response of the body. This is appropriate during stressful or hyperactive states of the body (i.e., exercise). Therefore, the sympathetic effect increases the following: rate of heartbeat, rate and depth of breathing, bloodflow to most skeletal muscles, and concentration of glucose in the blood. However, it also decreases certain processes: bloodflow to digestive organs and digestion.

The parasympathetic effect is the opposite of the sympathetic effect on an internal organ (e.g., decreases rate of heartbeat, increases digestion). This branch normally restores a normal, more relaxed state to body processes.

CHAPTER 9

The Sense Organs

9.1 Receptors

Receptors are specialized cells that can detect environmental changes called stimuli. Sense organs contain receptors. The skin, for example, is a sense organ that contains receptors that detect a wide variety of stimuli for touch, pressure, heat, cold, and pain.

Chemoreceptors detect chemical stimuli. These receptors are responsible for the senses of taste and olfaction. Each of the receptors for the four basic tastes (salty, sour, bitter, and sweet) are concentrated over different regions on the surface of the tongue. Different receptors in the epithelium of the upper nasal mucosa are responsible for the different senses of olfaction.

Proprioceptors are located in the joints of the body with tendons and ligaments. They are sensitive to stretching and pressure. They contribute to a person's knowledge of the position of body parts. The knee-jerk reflex begins with the stimulation of proprioceptors.

The retina of the eye contains **photoreceptors** (rods and cones) which are sensitive to light. The **mechanoreceptors** (sensory hair cells) of the inner ear are responsible for balance and hearing.

The development of any sense requires the activation of three components: (1) the stimulation of receptors, (2) impulse transmission by associated sensory neurons to the spinal cord and/or brain, and (3) interpretation of the sensory input by a mapped sensory area of the cerebral cortex.

The sense of vision develops by the activation of all three components. Retinal cells are stimulated. This activates sensory neurons of the optic nerve. Impulses along these neurons enter the brain and arrive at a mapped area on the cortex of the occipital lobe for interpretation.

9.2 Eye

The eye is a sensory organ with receptors located in its innermost layer, the retina. Several associated or accessory structure work with the eye for vision.

9.2.1 Associated Structures

Orbit - This is the eye socket. Seven different skull bones (e.g., frontal, maxilla, zygomatic) of the orbit are joined by sutures to house and protect most of the eye.

Extrinsic Muscles - Six skeletal muscles are attached to the eye. Four of the muscles have fibers ranging straight through the orbit (rectus = straight): **lateral rectus**, **medial rectus**, **superior rectus**, and **inferior rectus**. Two of the muscles have fibers ranging obliquely: **superior oblique** and **inferior oblique**. When each muscle contracts, it moves the eye in a specific direction. For example, the eye moves in a lateral direction when the lateral rectus contracts.

These muscles are controlled by motor neurons of cranial nerves III, IV, and VI (oculomotor, trochlear, and abducens).

Palpebra - The palpebra is the eyelid. A skeletal muscle inside the eyelid, the **levator palpebra**, contracts to raise the eyelid and open the eye. The **orbicularis oculi**, a series of circular muscle fibers around the front margin of the orbit, contract to oppose this action and close the eye.

The palpebra has an outer layer of skin. The inner surface is lined by the **conjunctiva**, a mucous membrane which doubles back over the exposed surface of the eye (cornea and part of the sclera). Eyelashes are attached to the edge of the palpebra.

Lacrimal Apparatus - The lacrimal apparatus is a series of structures that form, secrete, and drain the lacrimal fluid, (tear fluid). The **lacrimal gland**, embedded in the frontal bone above the eye, secretes the fluid through a group of **lacrimal ducts**. The fluid washes over the exposed surface of the eye, lubricating and moistening this surface. Two small ducts drain the fluid in the medial corner, passing it to the **nasolacrimal duct**. From this duct the fluid enters the **lacrimal sac**.

9.2.2 Structure of the Eye

Most of the makeup of the eye consists of several layers. The balance of the anatomy consists of three internal, transparent structures.

Sclera - Also known as the scleroid coat, this is the outer white covering of the eye. It provides protection much as the leathery cover protects a baseball.

Cornea - The sclera is modified anteriorly into a clear, centrally-located layer called the cornea. This convex layer is the first structure to intercept light rays. Its convexity allows it to refract (bend) these rays, concentrating them as they are collected to fall on the retina inside.

Choroid Coat - This dark, middle layer is highly vascular. The blood vessels deliver nutrients and oxygen to the tissues of the eye. The choroid coat is modified into three structures anteriorly.

Ciliary Body/Suspensory Ligaments - The choroid coat thickens anteriorly into a ring of smooth muscle, the ciliary body. Suspensory ligaments continue from it, attaching to the lens. The lens is a convex structure which is also elastic. Its convexity is determined by the degree of tension placed on it by the ciliary body and attached ligaments. The convexity will change for viewing objects at varying distances.

Iris - The iris is the third anterior modification of the middle layer. Anterior to the ciliary body, it is a ring consisting of two layers of smooth muscle, circular and radial, that surround an opening called the **pupil.** If a viewed area becomes brighter, contraction of the circular fibers constrict the pupil. If the area becomes darker, contraction of the radial fibers dilates the pupil. Therefore, the size of the pupil controls the amount of light passing through the eye.

Retina - The retina is the inner lining, covering about the posterior three-quarters of the eye. It does not have any anterior modifications. The retina contains two kinds of photoreceptors, **rods** and **cones**.

The cones are active in bright light and are responsible for color discrimination and the formation of sharp images. Color vision depends on the interaction of cones with three kinds of pigments—red, blue, and green. The concentration of cones is greatest in the **fovea centralis** (focal point) on the retina.

Rods contain rhodopsin, a pigment that becomes active during dim-light conditions. Their concentration increases with increasing distance from the focal point, just the opposite of the pattern for the cones. The rods are responsible for black and white vision.

The **optic disc** (blind spot) is a retinal region somewhat off center from the location of the focal point. Photoreceptors are absent here, as there is an opening for the passage of neurons of the **optic nerve.**

Lens - The lens is one of several transparent structures. It is a biconvex, elastic structure connected to the suspensory ligaments. Its convexity allows it to refract light rays.

Aqueous Humor - This is a fluid filling the anterior chamber of the eye. It is constantly being formed from capillaries in the ciliary body. It is drained by a circular canal (canal of Schlemm) where the sclera meets the cornea.

Vitreous Body (Humor) - This is a gelatin-like substance filling the larger posterior chamber. It maintains the shape of the eyeball. Its amount remains stable. Unlike the aqueous humor, it is not constantly drained and reformed.

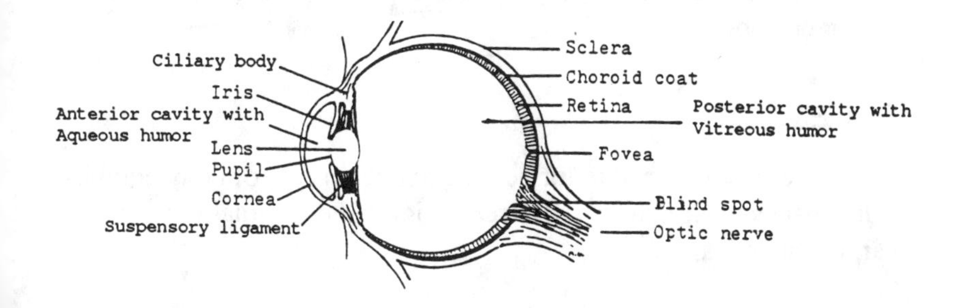

Figure 9.1 The Eye

9.2.3 Physiology of Vision

Light passes through the following sequence of structures before falling on the retina: **cornea - aqueous humor - lens - vitreous body - retina**. Each of these structures has the ability to refract light rays. Because of their convex shape, the cornea and lens bend the rays by causing their convergence. This is necessary, as light rays tend to scatter and diverge from any point on a viewed object. By convergence the light rays are collected and concentrated.

The majority of rays from any point of an object fall on or near the fovea centralis of the retina. The maximum concentration of cones is found here for maximum sensory ability and sharp image formation.

The convexity of the lens will change for viewing objects at varying distances. The pattern of response is:

closer object - more convexity - more convergence of rays
more distant object - less convexity - less convergence

Light rays diverge more from closer objects. Therefore, the convexity and refractive power of the lens must increase. For more distant objects light rays do not diverge as much. In this case the refractive power of the lens does not need to be as great and it flattens out somewhat, becoming less convex. The change in the shape of the lens for viewing objects of varying distances is called **accomodation**.

9.3 Ear

The ear is responsible for hearing and two types of body equilibrium (balance). It consists of three regions: the external ear, middle ear, and inner ear.

9.3.1 Structure

External Ear - The external ear consists of the **pinna (auricle)** and **external auditory meatus** (auditory canal). The pinna is the outer ear flap, consisting of elastic cartilage covered with skin. The meatus is a tube passing from the pinna into the temporal bone. Vibrating air waves (sound waves) pass through here as they travel toward the middle ear.

Middle Ear - The middle ear begins at the **tympanic membrane** (eardrum). The **malleus** (hammer) is a middle ear bone attached to this membrane. If the tympanum vibrates from the arrival and impact of air waves, the malleus transmits the vibrations to the **incus** (anvil) and **stapes** (stirrups). The stapes is attached to the **oval window** of the inner ear. The vibration of the stapes transmits a signal to this membrane.

The **eustachian tube** is a passageway extending from the middle ear to the nasopharynx, which is part of the throat. Air moved through this tube, by yawning or swallowing, can restore a pressure balance on the tympanum. The eustachian tube, however, is not involved in the transmission of vibrations for hearing.

Inner Ear - The inner ear is a complex series of channels. The **cochlea** is the part containing the receptors for hearing. It contains three channels: the **vestibular canal** (scala vestibuli), **tympanic canal** (scala tympani), and **cochlear duct** (canal). The vestibular and tympanic canals are connected at their ends.

As vibrations arrive at the oval window from the middle ear, they continue through **perilymph**, a fluid in the vestibular canal and tympanic canal. From the tympanic canal, vibrations are sent to the **round window**. This membrane is the last structure to receive vibrations, serving as a shock absorber.

As vibrations are passing through the perilymph, they disturb fluid, called **endolymph,** in the nearby cochlear duct. The cochlear duct contains the receptors, sensory hair cells, for hearing.

Two other structures of the inner ear, the **vestibule** and **semicircular** (half circle) **canals**, are involved with body equilibrium. The vestibule contains two small chambers, the **utricle** and **saccule**, that contain the receptors for static equilibrium. This involves balance when the body is relatively motionless. The three semicircular canals contain the receptors for dynamic equilibrium, maintaining balance when moving.

9.3.2 Physiology of Hearing and Equilibrium

The original stimuli for hearing are vibrating sound waves. They pass through the structures of the ear in the following order: **auricle** - **external auditory meatus** - **tympanum** - **malleus** - **incus** - **stapes** - **oval window** - **vestibular canal** - **tympanic canal** - **round window**. As each of these structures vibrates, they normally maintain the frequency (pitch) of the original sound waves in the air. The three bones of the middle ear amplify the vibrations, causing them to sound louder.

As the perilymph in the vestibular and tympanic canals transmits the vibrations, it disturbs the endolymph and sensory hair cells in the cochlear duct. The hair cells are receptors, mounted on a **basilar membrane** and suspended in the endolymph. As the cells rub against the nearby **tectorial membrane**, a stimulus is produced. The sensory hair cells, plus the basilar and tectorial membranes, compose the **organ of Corti**.

Action potentials from the sensory neurons of the **cochlear nerve** of cranial nerve VIII (vestibulocochlear) are transmitted to the temporal lobe of the cerebral cortex, where they are interpreted as sound.

For static equilibrium, the saccule and utricle are lined with sensory hair cells suspended in a gelatinlike matrix. Small deposits of calcium carbonate, called otoliths, contact different groups of hair cells depending on the position of the body. These stimuli are conducted to the brain for interpretation.

For dynamic equilibrium, sensory hair cells line the ampullae (expanded regions) where the semicircular canals meet the vestibule. They contact a gelatinlike matrix inside. There are three canals on each side of the body. The canals are at right angles to each other. Based on the changing direction and position of the body when moving, different groups of hair cells are affected in different combinations. This information is conducted to the brain.

Sensory neurons of the **vestibular nerve** transmit signals from the vestibule and semicircular canals. Combined with the cochlear nerve fibers for hearing, they form the vestibulocochlear nerve. The cerebellum is one brain region receiving an input for body balance.

CHAPTER 10

The Endocrine System

10.1 Hormone Action

The endocrine system consists of a group of ductless glands that secrete chemical signals, **hormones**, into the bloodstream. As a hormone is transported by the circulating blood, it signals target tissues that respond to this signal. For example, the posterior lobe of the pituitary gland secretes the hormone ADH (antidiuretic hormone). Transported by the blood flow, it reaches the kidney where its target tissues are found. At this structure it controls water balance.

Chemically, most hormones are either peptides (chains of amino acids) or steroids. Peptide hormones (i.e., insulin) bind to receptors on the cell surfaces at target tissues. This stimulates changes in the metabolism of the cells. One example is an increase in the rate of protein synthesis. Steroid hormones (i.e., aldosterone) bind to receptor molecules in the cytoplasm of the cell that enter the nucleus. In the nucleus they change the genetic activity of the cell.

Hormone signaling influences a wide variety of body functions, ranging from the growth of bones and muscles to the concentration of glucose in the blood.

10.2 Endocrine Glands

The main endocrine glands are the **pituitary** (anterior and posterior lobes), **thyroid**, **parathyroids**, **adrenal** (cortex and medulla), **pancreas**, and **gonads**.

The pituitary gland is attached to the hypothalamus of the lower forebrain.

The thyroid gland consists of two lateral masses, connected by a cross-bridge, that are attached to the trachea. They are slightly inferior to the larynx.

The parathyroids are four masses of tissue, two embedded posteriorly in each lateral mass of the thyroid gland.

One adrenal gland is located on top of each kidney. The cortex is the outer layer of the adrenal gland. The medulla is the inner core.

The pancreas is along the lower curvature of the stomach, close to where it meets the first region of the small intestine, the duodenum.

The gonads are found in the pelvic cavity.

10.3 Pituitary Gland

The pituitary gland consists of the anterior lobe (adenohypophysis) and posterior lobe (neurohypophysis).

Anterior Lobe - Small blood vessels continue from the **hypothalamus** to the endocrine cells of the anterior lobe. **Releasing factors**, chemical signals from the hypothalamus, reach the anterior lobe by the blood flow and influence its activity. Each releasing factor affects the secretion of a hormone from the anterior lobe. For example, **CRF** (cortico-releasing factor) signals the anterior lobe to secrete the hormone ACTH (adrenocorticotropic hormone).

Many of the hormones of the anterior lobe are **tropic hormones.** They signal target tissues in other endocrine glands. Tropic hormones include:

ACTH - signals the adrenal cortex

TSH - signals the thyroid gland

gonadotropic hormones - **FSH** in the male and female signals sex cell production and maturation. **LH** in the female signals ovulation. **ICSH** in the male signals the production of testosterone.

There are three related levels of signaling involving the hypothalamus, anterior lobe of the pituitary, and other endocrine glands. One example is:

I: hypothalamus secretes a releasing factor - i.e., CRF

II: anterior lobe of the pituitary secretes a tropic hormone - i.e., ACTH

III: adrenal cortex secretes cortisol

This response occurs when the level of sugar in the blood decreases. This change is sensed by the hypothalamus. By the three-tiered signaling through the hypothalamus - anterior lobe - adrenal cortex, the hormone cortisol stimulates a response that increases the level of sugar in the blood. This response reverses the original trend of a decreasing level of sugar in the blood. This pattern of response is an example of **negative feedback**—a response that reverses the trend of the original stimulus.

Some of the hormones secreted by the anterior lobe of the pituitary are not tropic hormones, for they do not signal other endocrine glands. They include:

Growth hormone (GH, somatotropin) - This stimulates the increase in the use of amino acids, particularly by bones and muscles. During the early years of physical growth, an oversecretion of GH can produce **gigantism**. A deficiency of GH, another imbalance during the years of physical growth, can produce **dwarfism**. An oversecretion of GH later in life produces a condition called **acromegaly**. Only certain bones (i.e., mandible) of the body are affected, leading to the disproportionate growth of body regions.

The growth hormone can also signal many body cells to use fats as a source of energy in preference to glucose. Glucose stays in the blood instead of entering cells as an energy source. Therefore, the growth hormone promotes a hyperglycemic (high blood sugar) effect.

Prolactin - This hormone stimulates the development of mammary glands to produce milk.

MSH (melanocyte stimulating hormone) - This hormone may stimulate the skin to increase the production of melanin, a dark pigment found in the skin.

Posterior Lobe - Axons from neurons in the hypothalamus continue into the posterior lobe. The two signals secreted from the posterior lobe are produced by these neurons.

ADH (antidiuretic hormone) - It signals the tubules in the nephrons of the kidney to reabsorb more water. First water enters the tubules by filtration. By reabsorption it is returned to the blood and, therefore, not eliminated. If the blood lacks sufficient water, the secretion of ADH increases, signaling the kidney to reabsorb more. If the blood has sufficient water, the secretion of ADH and rate of reabsorption decrease. ADH is also called **vasopressin,** as it can constrict blood vessels.

A deficiency of ADH leads to **diabetes insipidus**. Losing some ability to reabsorb water, the person eliminates a large amount of dilute urine.

Oxytocin - It signals the uterus to contract, inducing labor. It also stimulates the release of milk from the breasts when nursing.

10.4 Thyroid Gland

The thyroid gland secretes two hormones, the **thyroid hormone** and **thyrocalcitonin**.

The thyroid hormone consists of two components, thyroxin and iodine. This hormone increases the rate of metabolism of most body cells. A deficiency of iodine in the diet leads to the enlargement of the thyroid gland, known as a **simple goiter**. Hypothyroidism during early development leads to **cretinism**. In adults, it produces **myxedema,** characterized by obesity and lethargy. Hyperthyroidism leads to a condition called **exothalmic goiter**, characterized by weight loss as well as hyperactive and irritable behavior.

Thyrocalcitonin (calcitonin) decreases the concentration of calcium in the blood. Most of the calcium removed from the blood is stored in the bones.

10.5 Parathyroid Glands

The four parathyroids secrete the **parathyroid hormone** (PTH). It opposes the effect of thyrocalcitonin. It does this by removing calcium from its storage sites in bones, releasing it into the bloodstream. It also signals the kidneys to reabsorb more of this mineral, transporting it into the blood. It also signals the small intestine to absorb more of this mineral, transporting it from the diet into the blood.

Calcium is important for many steps of body metabolism. Blood cannot clot without sufficient calcium. Skeletal muscles require this mineral in order to contract. A deficiency of PTH can lead to **tetany,** muscle weakness due to a lack of available calcium in the blood.

10.6 Adrenal Glands

The adrenal **cortex** secretes at least two families of hormones, the **glucocorticoids** and **mineralcorticoids**. The adrenal **medulla** secretes the hormones **epinephrine** (adrenalin) and **norepinephrine** (noradrenalin).

Cortisol is one of the most active glucocorticoids. It generally reduces the effects of inflammation (i.e., swelling) throughout the body. It also stimulates the production of glucose for the blood from fats and proteins. This process is called **gluconeogenesis**.

Aldosterone is one example of a mineralcorticoid. It signals the tubules in the kidney nephrons to reabsorb sodium while secreting (eliminating) potassium. If sodium levels are low in the blood, the kidney secretes more **renin**, an enzyme that stimulates the formation of **angiotensin** from a molecule made from the liver. Angiotensin stimulates aldosterone secretion. As a result, more sodium is reabsorbed as it enters the blood.

The renin-angiotensin-aldosterone mechanism can raise blood pressure if it tends to drop. It does this two ways. Angiotensin is a vasoconstrictor, decreasing the diameter of blood vessels. As vessels constrict, blood pressure increases. In addition, as sodium is reabsorbed, the blood passing through the kidney becomes more hypertonic. Water follows the sodium into the hypertonic blood by osmosis. This increases the amount of volume in the blood and also increases blood pressure.

An oversecretion of the glucocorticoids causes **Cushing's syndrome**, characterized by muscle atrophy (degeneration) and hypertension (high blood pressure). A deficiency of these substances produces **Addison's disease**, characterized by low blood pressure and the development of stress.

Epinephrine and norepinephrine produce the "fight or flight" response, similar to the effect from the sympathetic nervous system. Therefore, they increase heart rate, breathing rate, blood flow to most skeletal muscles, and the concentration of glucose in the blood. They decrease the blood flow to digestive organs and diminish most digestive processes.

10.7 Pancreas

The pancreas contains exocrine and endocrine cells. Groups of endocrine cells, the **islets of Langerhans**, secrete two hormones. The beta cells secrete **insulin**; the alpha cells secrete **glucagon**. The level of sugar in the blood depends on the opposing action of these two hormones.

Insulin decreases the concentration of glucose in the blood. Most of the glucose enters the cells of the liver and skeletal muscles. In these cells this monosaccharide is converted to the polysaccharide glycogen. Therefore, insulin promotes **glycogenesis**, glycogen formation.

Glucagon promotes **glycogenolysis**, stimulating the breakdown of glycogen into glucose for release into the blood.

Insulin deficiency leads to the development of **diabetes mellitus**, specifically **type I** (juvenile) diabetes. As the pancreas does not produce sufficient insulin, it is treated by insulin injections. In **type II** (maturity onset) diabetes, the pancreas does produce enough insulin but the target cells do not respond to it.

10.8 Gonads

The main hormones from the reproductive organs are:

Testosterone - This hormone is more prominent in males. It belongs to the family of androgens, which are steroid hormones producing masculinizing effects. Testosterone stimulates the development and functioning of the primary sex organs. It also stimulates the development and maintenance of secondary male characteristics, such as hair growth on the face and the deep pitch of the voice.

Estrogen - In females this hormone stimulates the development of the uterus and vagina. It is also responsible for the development and maintenance of secondary female characteristics, such as fat distribution throughout the body and the width of the pelvis.

Progesterone - In females this hormone also stimulates development of primary and secondary female characteristics. Known as the hormone of pregnancy, it is very prominent in the last 14 days of the female reproductive cycle, occurring after ovulation. One of its effects is to increase the thickness and development of the uterine lining for implantation of an embryo, if fertilization occurs.

10.9 Other Hormones and Endocrine Glands

Endocrinology is a rapidly expanding field, with new endocrine structures and hormones being discovered constantly. Other well-known hormones, in addition to the main ones recognized in this chapter, include:

Erythropoietin - Secreted from endocrine cells in the kidney, it stimulates erythropoiesis (red blood cell formation).

Gastrin - Secreted from endocrine cells in the stomach, it stimulates increased gastric secretions from this organ.

Secretin - Secreted from endocrine cells of the small intestine, it stimulates increased secretions from the pancreas into the small intestine.

Prostaglandins - This is a group of local hormones throughout the body that act close to their site of secretion. Their functions range from influences on blood clotting to effects on inflammation.

CHAPTER 11

The Circulatory System

11.1 Functions

The circulatory system consists of the **blood**, **heart**, and **blood vessels**. The heart functions as a pump, forcing blood through a series of vessels that distribute the blood to the cells of all body regions. The blood offers oxygen and substances necessary for cell metabolism. The blood also removes waste products from these cells. As the blood carries out this role of internal transport, it travels to and from the various body regions it serves. In other words, it circulates.

The circulatory system meets other needs of body metabolism. Buffers in the blood are substances that stabilize the pH of the fluid surrounding cells (extracellular fluid). By the dilation and constriction of blood vessels in the skin, heat from the body can be either liberated or conserved. This response contributes to the control of internal body temperature.

The circulatory system is also part of the immune system of the body. White blood cells (leukocytes) act as a main line of defense that fight infection. In addition, the circulating blood is a vehicle for transporting hormones from endocrine glands to target tissues.

11.2 Blood

The adult human body contains an average of five liters of blood. This type of connective tissue consists of three kinds of specialized cells (formed elements): **erythrocytes** (red blood cells), **leukocytes** (white blood cells), and **thrombocytes** (platelets). As the blood circulates these cells are suspended in a liquid matrix, the **plasma**. Normally, the cells make up about 45% of the blood by volume. The blood can also be typed into four classes (ABO bloodtyping).

The percentage of the blood that is cellular by volume is called the **hematocrit**. This can be measured and computed by laboratory tests. The cells in a test-tube sample of blood can be packed into the bottom of the tube. The plasma, which is not as dense as the cells, settles above the cells in the test tube. If, for example, the cells make up 9 volume units in a total of 20 units of blood, the hematocrit is 45% ($9/20$). The plasma in this case composes the other 55% ($11/20$).

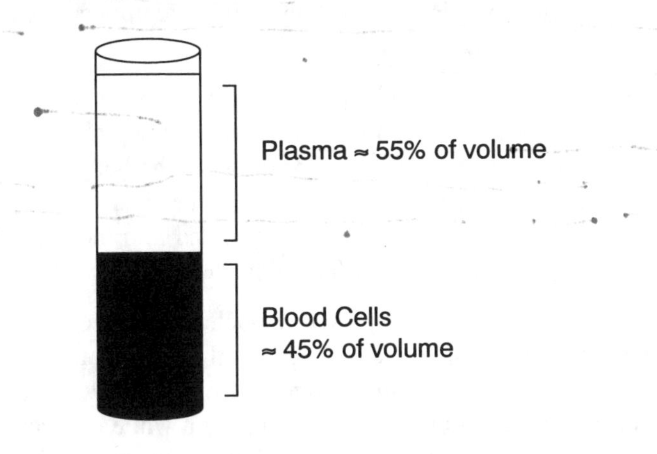

Figure 11.1 Test Tube of Blood – Cells and Plasma

Erythrocytes - Erythrocytes are the gas carriers of the blood. They are small, biconcave disks. At maturity they lack a nucleus and the organelles common to most cells. They do contain millions of molecules of **hemoglobin**. This pigment has the ability to combine reversibly with oxygen. Oxygen combines with hemoglobin, forming oxyhemoglobin, as blood passes through the lungs. It dissociates from hemoglobin at tissue sites throughout the body, releasing it for use by cell metabolism.

The hemoglobin in erythrocytes also transports carbon dioxide. It receives carbon dioxide from cells as a product of their metabolism, then unloads it at the lungs for elimination from the body.

Erythrocytes, along with leukocytes and thrombocytes, can be counted on a per cubic millimeter (mm) basis. A normal count ranges from about 4.2 million to 6 million per cubic mm. The count is usually somewhat higher in males than in females. An abnormally low red cell count and/or hemoglobin deficiency produces an **anemia,** the inability of the blood to carry and deliver sufficient oxygen to the body cells.

Erythrocytes are produced from the red bone marrow in the ends of long bones, cranial bones, ribs, sternum, and vertebrae. The early stages of red cells have a nucleus. Lacking a nucleus by the time they are released into the circulation, they cannot reproduce. A normal lifespan for a red blood cell is 120 days.

Eventually red blood cells are destroyed by phagocytic cells in the liver and spleen. The hemo portion of the hemoglobin molecule, containing iron, is sent to the liver for reuse by the body. The globin portion, a protein, is converted into bile in the liver by a series of steps. Stored in the gallbladder, bile is secreted into the small intestine for the physical digestion of fats in the diet.

Leukocytes - Leukocytes have a nucleus and are larger than erythrocytes. Most of them are produced in the bone marrow. A **total** white blood cell **count** normally ranges from 5,000 to 10,000 per cubic millimeter. **Leukopenia** is revealed by a white cell count that is less than the lower end of this range. Leukocytosis is a high white cell count, above 10,000.

There are five different kinds of white blood cells. Each kind makes up a percentage of the total white cell count. White blood cells lack pigments. They can be stained and examined under the microscope in order to be distinguished from each other. Each kind of white cell has unique characteristics revealed through this procedure.

Three kinds of leukocytes contain stained granules in the cytoplasm: **neutrophil, basophil,** and **eosinophil**. Two kinds lack granules: **lymphocyte** and **monocyte**. The neutrophil, for example, has a nucleus with many interconnected lobes and blue granules. By contrast, the nucleus of the eosinophil lacks lobes but it has reddish-orange granules.

In addition to the total count, a **differential** white cell **count** can also be conducted on white blood cells. This charts the percentage of each kind of cell in the total. Neutrophils, for example, make up around 65%. Lymphocytes make up about 25%. Information from a differential count can have diagnostic value. If, for example, the percentage of eosinophils increases, it usually means the body has an allergy.

Leukocytes fight infection in several ways. Some are phagocytic, capable of engulfing foreign cells such as invading bacteria. They can leave the circulation and enter the intercellular spaces to do this, an action called **diapedesis.** Lymphocytes can produce molecules, called **antibodies**, that react against foreign substances entering the body, **antigens**, reacting with them and destroying them.

Thrombocytes - The platelets are cell fragments, produced from large cells, megakaryocytes, in the bone marrow. A normal platelet count ranges from 200,000 to 400,000 per cubic millimeter.

Platelets initiate the process of blood clotting. They are attracted to a rough texture on the inside surface of a blood vessel. Normally this surface is smooth, but can be changed by a cut or rough deposit (i.e., cholesterol).

If platelets are attracted to a rough surface, platelets release a group of substances called platelet factors (thromboplastins). These factors react through a series of chemical changes, leading to the conversion of fibrinogen to **fibrin**. Fibrinogen is a plasma protein. Fibrin is an insoluble protein that traps red blood cells, forming a clot. This clot can plug vessels, preventing blood from escaping.

Plasma - As the liquid portion of the blood, the plasma is 90 to 92 percent water. Many kinds of substances are dissolved or suspended in this medium. Some are nutrients from the diet: glucose and amino acids. Others are waste products: lactic acid and urea (for nitrogen elimination). Other substances include minerals, hormones, and antibodies. All of these substances are part of body metabolism.

Blood Typing - The most common typing system of the blood is the **ABO** system, depending on the presence or absence of two **antigens**, A and B. If only one is present on the surface of the erythrocytes, the bloodtype is either **A** or **B**. If both are present, the bloodtype is **AB**. If neither is present, the type is **O**, the most common ABO bloodtype. Therefore, the antigen content names the bloodtype.

Two **antibodies**, anti-A and anti-B, are present in the plasma and are also part of the ABO system. This antibody makeup is normally the opposite, by letter, of the antigen in a person's blood. For example, if a person has only the A antigen (type A), anti-B is present in the person's plasma. A type B person (B antigen) has the anti-A antibody. A type AB person (both antigens) has neither antibody. A type O person has both antibodies.

An **agglutination** (clotting) reaction occurs if the antigen and antibody of the same letter are mixed. Although this does not normally occur in a person's own blood, it can occur through a mistake of blood transfusion. If type A blood is transfused into a person with type B blood (anti-A antibodies present), the two bloodtypes will clot, leading to the death of the individual receiving the blood.

11.3 Heart – Structure and Function

The heart works as a double pump. The right side pumps blood to the lungs; the left side sends the blood to all other body locations. The adult heart is about the size of a closed fist. About two-thirds of its mass is to the left of the body midline. Residing in the mediastinum, the heart is housed in a loose-fitting sac, the **pericardium**. Its apex, the cone-shaped inferior portion, fits into a depression on the diaphragm.

The anatomy of the heart consists of chambers, associated blood vessels, valves, and a conduction system.

Chambers - The **atria** (sing., atrium) are the two superior chambers of the heart. A **ventricle** is inferior to each atrium on each side of the heart. The ventricles are larger chambers with thicker muscular walls (cardiac muscle tissue). The left ventricle is the most powerful pumping chamber of the heart. A septum (wall) separates the right atrium from the left atrium. It also separates the two ventricles.

The muscle wall of each chamber, the **myocardium**, is lined internally by the **endocardium**, a layer of epithelium.

Associated Blood Vessels - Several major blood vessels connect to the heart, transporting blood to and from this organ. Each vessel is either an artery or vein. **Arteries** transport blood away from the heart. **Veins** return blood to the heart.

The **superior** and **inferior vena cava** are the two thickest veins of the body. The superior vena cava drains blood from the head, neck, and arms into the right atrium. As the second longest vein of the body, the inferior vena cava transports blood into the right atrium from all other body regions.

The **pulmonary trunk** transports blood from the right ventricle. Dividing into a right and left **pulmonary artery**, the blood flow continues to the lungs.

Four **pulmonary veins**, two from each lung, return blood to the left atrium.

The **aorta**, the largest artery in the body, transports blood from the left ventricle, priming the arteries that serve all body regions except the lungs. Shortly after the aorta exits from the left ventricle, two **coronary arteries** send some blood back to the heart tissues. This blood supplies the heart cells with their need for oxygen and nutrients.

Valves - There are two pairs of valves in the heart. The **AV** (atrioventricular) **valves** are found between the atria and ventricles. Also called the cuspid valves, the **bicuspid** valve (mitral valve) is between the atrium and ventricle on the left. The **tricuspid** valve is between these chambers on the right.

The semilunar valves (half-moon shaped) are located where the major artery exits from the ventricle on each side. The **pulmonary semilunar valve** is located where the pulmonary trunk exits from the right ventricle. The **aortic semilunar valve** is located where the aorta exits from the left ventricle.

Valves can open and close. When closed, they prevent the backflow of blood. On each side the blood flows from the atrium to the ventricle through the outgoing artery. The AV (cuspid) valves close to prevent the backflow of blood to the atria. Each AV valve is attached to the **chordae tendinae**. These stringlike structures are anchored to the **papillary muscles**, elevated masses of muscle tissue in the ventricles.

The semilunar valves close to prevent the backflow of blood to the ventricles, sealing off this possibility and assuring that blood flows through the pulmonary trunk and aorta instead.

Conduction System - The heart contains masses of nodal tissue, excitable tissue that conducts impulses and stimulates the heartbeat intrinsically. This conduction system signals the heart to beat independently. It does not require any external influences. The impulse to stimulate the heartbeat passes through the conduction system structures in this order: **SA node** - **AV node** - **AV bundle** - **Purkinje fibers**.

The SA node, known as the pacemaker, is in the wall of the right atrium, near the entrance of the superior vena cava. By initiating the impulse for each heartbeat, the SA node sends an impulse through the myocardium of both atria, depolarizing the muscle cells of these chambers. This causes atrial contraction.

The AV node at the base of the right atrium receives an impulse from the walls of the atrium, relaying it through the left and right bundle branches of the AV bundle (bundle of His), located in the septurn between the two ventricles. At the apex of the heart, these two branches subdivide into numerous Purkinje fibers. These fibers deliver impulses to the myocardial cells in the ventricles, depolarizing them for contraction.

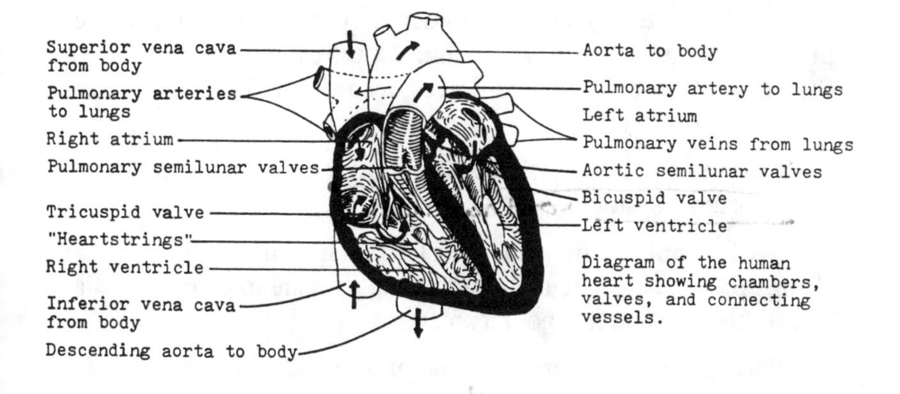

Figure 11.2 Anatomy of the Heart

11.3.1 Cardiac Cycle

The cardiac cycle is a repeating series of events that take place each time the heart beats. These events include the **systole** (contraction) and **diastole** (relaxation) of the heart chambers. The two sides of the heart do not contract independently but respond in unison. If the cycle requires 0.8 second (s.), its events can be summarized over three steps.

I: atria - systole, ventricles - diastole (first 0.1 s.)

II: atria - diastole, ventricles - systole (next 0.3 s.)

III: atria - diastole, ventricles - diastole (last 0.4 s.)

First, as the atria contract, the ventricles relax. As the ventricles contract, the atria relax. For the last half of the cycle, all chambers are relaxed.

The **cardiac output** (CO) is the total volume of blood pumped from the left ventricle through the aorta each minute. It is the product of the **heart rate** and **stroke volume**. For example, if the heart beats 70 times per minute and ejects 80 ml of blood per beat (stroke volume), the cardiac output is:

70 × 80 ml = 5,600 ml or 5.6 liters

The **ECG** (EKG), electrocardiogram, is a recording of the electrical activity of the heart, accounting for the events of the cycle. The ECG consists of the following waves, recorded in this order for each cardiac cycle:

P wave - from depolarization of the atria—this electrical change produces atrial systole.

QRS complex - from the depolarization of the ventricles—this change produces ventricular systole. The atria are experiencing diastole at this time, but it is not recorded.

T wave - from the repolarization of the ventricles—this produces ventricular diastole

11.4 Blood Vessels – Structure and Function

Arteries are vessels that conduct blood away from the heart. They branch into arterioles, vessels that are smaller and more numerous. Veins are vessels that transport blood to the heart. They form by the collection of venules, vessels that are smaller and more numerous. To supply any region of the body, the blood circulates through the following sequence of vessels: **arteries - arterioles - capillaries - venules - veins.**

Capillaries are microscopic vessels that are very numerous. Their function is exchange, meaning a two-way transport.

The **systemic circulation** contains this sequence of vessels. Originating with the aorta, the blood in this circuit contains blood high in oxygen and low in carbon dioxide. As blood circulates systemically, oxygen and nutrients leave the blood of the capillaries and enter body cells. Carbon dioxide and waste products leave cells and enter the blood. The systemic blood returns by veins to the right atrium.

The **pulmonary circulation** is a shorter route, beginning with the pulmonary trunk connected to the right ventricle. It contains blood low in oxygen and high in carbon dioxide. This situation is reversed at the capillaries in the lungs. This circuit ends where the pulmonary veins enter the left atrium. The blood now has more oxygen and less carbon dioxide. As it passes through the left side of the heart, it enters the aorta and becomes part of the systemic circulation.

The wall of a capillary consists of one layer of flat epithelial cells (simple squamous epithelium). This is suitable for their exchange role. The walls of all other vessels contain three layers.

Fibrous connective tissue composes the outer layer of an artery or arteriole, the **tunica externa**.

The middle layer, the **tunica media**, consists of smooth muscle with elastic fibers. Contraction of the muscle tissue produces the constriction of the vessel, called vasoconstriction. Relaxation of this tissue produces vasodilation, an increase in the diameter of the vessel.

The inner layer, the **tunica intima**, consists of simple squamous epithelium (endothelium). It provides a smooth surface for the passage of blood through the vessel.

If viewed by a transverse section, a vein or venule has the same three layers. However, the middle layer is proprotionately thinner. Also, the inner layer has extensions called **valves.** Valves prevent the backflow of blood. Therefore, the blood continues to flow toward the heart.

Arteries and arterioles are **elastic** and **contractile**, due to the thickness of their middle layer. The blood pressure in systemic arteries is high, as they are near the left ventricle which influences this pressure by contraction. The pressure in these arteries is also **pulsatile**, meaning it fluctuates due to the systole and diastole of the left ventricle during the cardiac cycle. When blood arrives in the arterioles, the pressure is lower and not pulsatile.

Blood pressure, which is the same in all large systemic arteries throughout the body, can be measured by two numbers expressed in millimeters (mm) of mercury. The higher number is the **systolic** pressure, produced during the contraction (systole) of the left ventricle. When this chamber is relaxed (diastole) during the last half of the cardiac cycle, the pressure drops to a lower number, the **diastolic** pressure.

If, for example, the blood pressure is 120/80, 120 is the systolic pressure, 80 is the diastolic pressure.

The **pulse pressure** is the difference in these two numbers. In the previous case it is 40 mm.

Veins and venules have the opposite properties of arteries. They are not as elastic or contractile, due to a thinner middle layer. The blood pressure in systemic veins is much lower (i.e., 10 to 30 mm), as they are near the end of the circulatory pathway that returns blood to the right side of the heart. The pressure is not pulsatile. Whereas a **pulse** (pulse pressure) can be felt in an artery (i.e., radial artery in the wrist), it cannot be found in a vein.

Blood circulates from arteries to veins due to a pressure difference, moving continuously from regions of higher pressure to regions of lower pressure. From arteries and arterioles, blood arrives at the capillaries.

When blood enters the systemic capillaries, the blood pressure is sufficient to force some plasma out into the intercellular spaces. Here it becomes **tissue fluid**. This fluid contains substances (i.e., oxygen) to supply body cells.

At the opposite end of the capillaries, the blood pressure has dropped significantly. It is less than an osmotic pressure that draws the fluid back into the capillaries. The osmotic pressure is due to the presence of plasma proteins in the blood. They make the blood hypertonic, causing it to draw fluid from the tissue spaces by osmosis. Therefore, most of the tissue fluid returns to the capillaries at this end.

The tissue fluid entering the blood in the capillaries carries waste products from body cells. This blood enters the systemic venules and veins.

11.4.1 Circulation – Arteries and Veins

Arteries - All major arteries subdivide into smaller arteries that form arterioles. These arterioles deliver the blood to millions of capillaries that supply body regions.

The **aorta** is the largest artery. Beginning as the ascending aorta from the left ventricle, it forms an **arch**. Three major arteries distribute blood from the arch, delivering it to the head and arms. The **left subclavian** artery is lateral to the **left common carotid** artery. On the right side of the arch, the **brachiocephalic** (innominate) artery leaving the arch splits into the **right common carotid** and **right subclavian** arteries.

The common carotid on each side subdivides into an external and internal carotid artery, which supply the head and neck.

The subclavian artery on each side supplies the arm by the following series of arteries: **axillary** - **brachial** - **radial** - **ulnar**. Blood flows through them in a distal direction. The subclavian becomes the axillary at the armpit. The axillary becomes the brachial as it parallels the humerus. In the forearm blood enters the radial artery laterally and the ulnar artery medially.

Beyond the aortic arch, the aorta descends posterior to the heart. It continues to transport blood through the thorax and abdomen. Some arteries that branch from the **descending aorta** are:

celiac - supplies the stomach, liver, and spleen

superior mesenteric - mainly supplies the small intestine

renal - a pair of arteries supplying the kidneys

inferior mesenteric - supplying most of the large intestine

At the level of the ilium of each hip bone, the aorta subdivides into two **common iliac** arteries. Each common iliac artery forms an **external** and **internal** artery. The internal iliac passes medially into the pelvic cavity, supplying the urinary and reproductive organs.

Each external iliac artery supplies the leg by the following series of arteries: **femoral** - **popliteal** - **anterior** – **posterior tibial**. The blood flows through them in a distal direction. The femoral artery begins in the upper three-fourths of the thigh. It has a branch, the **deep femoral**. The femoral becomes the popliteal in the posterior region of the knee. From the popliteal, blood flows into the two tibial arteries in the front and back of the shinbone.

Veins - Blood from the systemic capillaries collects into venules which transport the blood into larger and larger veins as it returns to the heart. The **superior** and **inferior vena cava** are the two thickest veins of the body, returning blood to the right atrium.

The head and arms are drained by four veins—two **internal**

jugular veins that are larger and medial, plus two **external jugular** veins that are smaller and lateral. Each internal jugular drains into a **brachiocephalic** (innominate) vein. These two veins form a V, merging to send blood into the superior vena cava. Each external jugular connects to a **subclavian** vein which drains the arm.

Distal to each subclavian vein, three major veins collect blood from each upper arm. The **cephalic** vein is lateral to the humerus. The **basilic** vein is medial. The **brachial** vein is posterior to the humerus. Near the proximal end of the humerus, the brachial and basilic veins form the **axillary** vein. The axillary vein sends blood to the subclavian vein. Laterally, the cephalic vein drains into the subclavian directly.

In the elbow region, some blood from the lateral forearm does not enter the cephalic vein but crosses over to the basilic through the **medial cubital** vein.

An **accessory cephalic** vein drains the lateral forearm, connecting to the cephalic vein at the elbow.

Two major venous routes drain blood from the leg. Laterally, the blood flows through this sequence of veins, moving in a proximal direction: **anterior and posterior tibial - popliteal - femoral - external iliac**. Medially, blood passes from the **dorsal venous arch** in the foot into the **great saphenous** vein, the longest vein of the body. The great saphenous vein connects to the femoral vein near the proximal end of the femur.

Each external iliac vein merges with an **internal iliac** vein, which is medial and drains the structures in the pelvic cavity. This merger forms the **common iliac** vein on each side. The two common iliac veins form the inferior vena cava. As the second longest vein, it transports blood through the abdomen and thorax, returning it to the right atrium.

Hepatic Portal System - Many organs in the abdominal cavity are drained by veins that deliver blood to the liver before it is released into the inferior vena cava for return to the heart. At the liver this blood passes through another series of capillaries. This hepatic portal circulation ranges from the capillaries of the following organs to the capillaries of the liver: pancreas, gallbladder, spleen, stomach, small intestine, and large intestine.

The liver controls the concentration of glucose, toxins, and other substances in the blood passing through it. After these adjustments are made, the capillaries in the liver collect into larger vessels, forming the hepatic vein. It transports blood into the inferior vena cava.

11.5 Lymphatic System

The lymphatic circulation is sometimes considered a separate system from the rest of the circulatory system. It begins with **lymph capillaries**, microscopic vessels that take up a product of the tissue fluid not recaptured by the nearby systemic capillaries. This fluid is called **lymph**. It contains more protein that normal tissue fluid.

The lymph capillaries collect into larger and larger vessels, the lymphatics. These lymphatics parallel the systemic venules and veins in body regions, but contain more valves. The final two collecting vessels are the **thoracic duct** and **right lymphatic duct**. The right lymphatic duct drains the right arm and right halves of the head and thorax. The thoracic duct drains the remainder of the body. Both of these vessels add lymph to the systemic blood in veins near the heart.

Lymph nodes are found at certain sites of the lymphatic circulation. The major lymph nodes are the cervical, axillary, cubital, and inguinal. The nodes contain phagocytic cells that filter the blood and fight infection. In addition, some lymphocytes are produced in the lymph nodes.

Lymph capillaries in the wall of the small intestine absorb fatty acids from the diet.

CHAPTER 12

The Respiratory System

12.1 Respiration

Respiration is a process that involves the distribution and use of gases, oxygen and carbon dioxide, by the body. It occurs at several different levels:

1. **breathing** - the inhalation and exhalation of air
2. **external respiration** - the exchange of gases between the lungs and the blood of the pulmonary capillaries
3. **internal respiration** - the exchange of gases between the blood of the systemic capillaries and cells

Inhaled and exhaled air passes through a series of chambers during inhalation and exhalation. These chambers are continuous and are structures of the respiratory tract. Inhaled air passes through the chambers of this tract in the following order:

nose - pharynx - larynx - trachea - primary bronchi - secondary bronchi - bronchioles - alveolar ducts - alveoli

12.1.1 Anatomy – Respiratory Tract

Nose - The external nose projects from the face. The larger internal nose is superior to the oral cavity, separated from it by the hard palate. Several bones of the skull (i.e., ethmoid) are the boundaries of the internal nose.

Inhaled air passes through a pair of **external nares** (nostrils) of the external nose. The mucous membranes lining the internal nose moisten and filter the air. The plentiful blood supply in these membranes also warms the air. Olfactory receptors in the roof of the internal nose detect different chemicals in the air, initiating the sense of olfaction (smell).

Pharynx - Inhaled air passes through a pair of **internal nares** in the posterior portion of the internal nose. From these openings air enters the pharynx (throat). This muscular tube, about five inches in length, is lined with mucous membranes.

Vertically, the throat consists of three regions: the **nasopharynx**, **oropharynx**, and **laryngopharynx**. The nasopharynx extends from the internal nares to the level of the soft palate of the oral cavity. The oropharynx extends from this palate to the hyoid bone. The laryngopharynx is the remaining inferior portion of the throat.

The throat has several pairs of tonsils that protect the body from invading bacteria. The pharyngeal tonsils (adenoids), for example, are in the posterior wall of the nasopharynx.

Larynx - The larynx (voicebox) consists of nine pieces of cartilage. The three largest pieces are the **epiglottis**, **thyroid cartilage**, and **cricoid cartilage**.

The epiglottis is the superior, leaflike part. It is attached to the larynx posteriorly, capable of covering its opening at the top, the **glottis**. It normally covers this opening when food is swallowed. The thyroid cartilage ("Adam's apple") is the large, triangular piece. The cricoid cartilage is the most inferior piece, attaching the larynx to the first ring of the trachea.

Folds of the mucous membranes lining the inside of the larynx, the **vocal cords**, extend across the interior of this structure. Skeletal muscles cause them to vibrate at different frequencies, determining the pitch of voice production.

Trachea - The trachea (windpipe) is a tube about five inches long and one inch in diameter. Its wall of smooth muscle is reinforced with rings of C-shaped cartilage. The open end of the C's are directed toward the esophagus of the digestive tract, which is posterior to the trachea. The trachea is lined with mucous membranes.

Primary Bronchus - The base of the trachea subdivides into a left and right primary (main) bronchus. This arrangement resembles an inverted Y, with each bronchus entering one of the lungs. Each lung is an elastic, baglike structure that houses the remainder of the respiratory tract. The left lung is smaller than the right lung.

Secondary Bronchus - Each primary bronchus subdivides into secondary bronchii. There are three secondary bronchii in the right lung, one serving each lobe of the lung (three lobes). There are two secondary bronchii in the left lung, one serving each lobe of this lung (two lobes).

Bronchioles - The tract continues to subdivide into smaller, more numerous passageways in the lung. This pattern resembles an inverted tree. Bronchioles are very small tubes, lacking the cartilage rings found in the bronchii.

Alveolar Ducts, Alveoli - Most of the interior of the lungs is filled with microscopic air sacs called alveoli (sing., alveolus). The alveoli exist as clusters. Each one in a group receives inhaled air from an alveolar duct. The pair of lungs have an estimated 300 million alveoli. The alveoli are close to the capillaries of the pulmonary circulation. The thin walls of the alveoli consist of one layer of cells, simple squamous epithelium.

Lungs - Each lung has a broad, inferior base that rests on the diaphragm. Its **apex** is the cone-shaped, superior portion, projecting about one inch above the clavicle. Most of the surfaces of the lung are costal, facing and contacting the ribs.

Each lung resides within an **intrapleural cavity**. This is a sealed-off space within two layers of serous membranes. The **parietal pleura** lines the inside surface of the chest wall. The **pulmonary pleura** adheres to the lung.

The pressure in the intrapleural cavity is normally negative, meaning that it is below atmospheric pressure. At standard conditions of sea level, atmospheric pressure is 760 mm of mercury when measured by a barometer. If, for example, the intrapleural pressure is 757 mm, it is three units below this (minus 3 compared to a standard of 0).

12.2 Breathing – Physiology

Breathing is a cycle involving the alternation of two processes, inhalation and exhalation. Each process unfolds through a series of steps, beginning with a pressure in the lungs of 0 (760 mm).

Inhalation - The inflow of air results from a series of volume and pressure changes.

1. The **respiratory center** in the medulla sends a signal to the **diaphragm** and **intercostal muscles**. This signal causes these skeletal muscles to contract.

2. When relaxed, the diaphragm is elevated. As it contracts, it flattens. As it contracts it increases the size of the intrapleural cavity vertically. The intercostal muscles also contract, elevating the ribs. This increases the size of the intrapleural cavity transversely.

3. As the size of the intrapleural cavity increases, the pressure inside of it decreases. This inverse relationship between volume and pressure is stated by **Boyle's law**. With a normal, quiet inhalation (tidal volume), the pressure in the cavity drops from minus 3 to minus 6.

4. The lungs are drawn into the region of lower pressure in the intrapleural cavity surrounding them. As their volume increases, the pressure inside them decreases. With a normal inhalation, the pressure in the lungs drops from 0 to –3.

5. The pressure in the lungs, particularly the alveoli, drops below atmospheric pressure. The pressure in the lungs is now –3 compared to 0 (760 mm) in the atmosphere.

6. Air is drawn from the region of higher pressure (atmosphere, 0) to the region of lower pressure (–3, lungs). As a result, air rushes into the lungs, producing an inhalation.

Exhalation - The outflow of air results from a series of volume and pressure changes that reverses the situation that produced an inhalation.

1. The diaphragm relaxes, returning to an elevated position. The intercostal muscles relax, causing the ribs to drop.
2. The relaxation of these skeletal muscles decreases the size of the intrapleural cavity.
3. As the size of the cavity decreases, the pressure in the cavity increases. With a normal exhalation, this increase is from about –6 to –3.
4. The lungs are compressed somewhat by the decrease in the volume of the intrapleural cavity around them. As their volume decreases, the pressure in them increases.
5. The pressure in the lungs is also increased by the recoil of the elastic tissue in their walls. This recoil causes the pressure in the lungs to become temporarily positive (i.e., +3 or +4).
6. Air is forced from the region of higher pressure (lungs, +3) to the region of lower pressure (atmosphere, 0). As a result, air moves out of the lungs, producing an exhalation.

When the positive pressure in the lungs drops back to 0, the cycle is completed and can begin again.

The volume of air that is inhaled or exhaled can vary depending on the metabolic needs of the body. Air volumes, measured through **spirometry** in the lab, include:

tidal volume (TV) - volume of air normally inhaled or exhaled by quiet breathing. One example is 500 ml.

expiratory reserve volume (ERV) - This is the amount of air that can be exhaled in addition to the tidal volume. It can be as much as 1,000 ml.

inspiratory reserve volume (IRV) - This is the amount of air that can be inhaled in addition to the tidal volume. In some individuals this can be as high as 3,000 ml.

vital capacity (VC) - This is the total lung capacity. It is the sum of the tidal volume plus the two reserves:

VC = TV + IRV + ERV

residual volume - This is a remaining volume of air in the lungs that cannot be forcibly expired. It can range from 800 to 1,000 ml.

The rate and depth (volume) of breathing changes to meet the changing requirements of body metabolism. The total volume of air inhaled or exhaled per minute can be computed by multiplying rate and depth. For example:

12 cycles/min × 500 ml = 6,000 ml

The respiratory center in the medulla will increase rate and depth if the level of carbon dioxide and hydrogen ions (acidity) increases in the arterial blood. This increase eliminates these waste products. Also, as the body breathes faster and deeper, more needed oxygen is acquired.

12.3 External Respiration

Blood transported from the right side of the heart has a low concentration of oxygen and a high concentration of carbon dioxide. As this blood arrives in the capillaries of the pulmonary circulation, these situations are reversed by the exchange processes of external respiration. The events of these processes are:

1. Oxygen diffuses from the alveoli into the blood.
2. Carbon dioxide diffuses from the blood into the alveoli.

The concentrations of gases are expressed by **partial pressures**. Each gas in a mixture of gases contributes its part to the entire gas pressure. For example, in the atmosphere, the partial pressure of oxygen is about 152 mm of mercury. This is because oxygen comprises about 20% of the total air pressure (0.2 × 760 = 152).

The partial pressure of oxygen in the alveoli is often around 100. In the blood of the pulmonary capillaries, it is much lower (i.e., 35). Therefore, oxygen diffuses from the alveoli into the blood, raising the pressure in the blood to about 100.

The partial pressure of carbon dioxide in the alveoli is low, perhaps around 40. In the blood of the pulmonary capillaries it is higher, at about 48. Therefore, carbon dioxide diffuses from the blood into the alveoli, dropping the pressure in the blood to about 40.

Diffusion is a transport process that tends toward an equilibrium. Therefore, each amount of blood returning to the left side of the heart has partial pressures that have changed to about:

oxygen: 100 (up from 35)
carbon dioxide: 40 (down from 48)

Each pressure represents the level in the alveolus. The blood has equilibrated with concentrations close to these levels.

12.4 Internal Respiration

The blood returns to the left side of the heart after external respiration. It is pumped through the aorta from the left ventricle, reaching tissue cells by the systemic circulation. In the systemic capillaries, the processes of internal respiration occur. These processes are:

1. Oxygen diffuses from the blood into the tissue cells.
2. Carbon dioxide diffuses from the tissue cells into the blood.

Oxygen is used as a reactant by cellular respiration. Cellular respiration also produces carbon dioxide, which is a waste product that must be eliminated from tissue cells.

The partial pressure of oxygen in the blood of the systemic capillaries is often about 100. In the tissue cells it is about 40. Therefore, oxygen diffuses into the cells.

The partial pressure of carbon dioxide in the blood of the systemic capillaries is about. 40. In the tissue cells it can be 48. Therefore, carbon dioxide diffuses into the blood.

By equilibration through diffusion, the partial pressures for the blood returning to the right side of the heart are about:

oxygen: 35 (down from 100)
carbon dioxide: 48 (up from 40)

For each amount of blood returning to the lungs, these values for partial pressures are changed again through the processes of external respiration. As a summary, one example of these changes is:

oxygen: 35 to 100 as the blood acquires this gas from the lungs by external respiration; 100 to 35 as the blood loses some of this gas to cells by internal respiration

carbon dioxide: 48 to 40 as the cells lose some of this gas by external respiration; 40 to 48 as the blood acquires this gas from the cells by internal respiration

CHAPTER 13

The Digestive System

13.1 Digestion

Most molecules in the human diet are too large to be used in metabolism. Starch, as one example, is a complex carbohydrate. This polysaccharide is too large to be absorbed into the bloodstream and transported by the circulation.

The digestive system prepares food molecules for use in the body. It accomplishes this through physical digestion and chemical digestion.

13.1.1 Physical Digestion

Part of the digestion of food involves a physical change. Masses of food are broken into smaller particles without altering the chemical structure of the molecules. Physical digestion begins in the oral cavity with mastication (chewing). Another example occurs in the small intestine where the secretion of bile emulsifies fat globules, subdividing them into smaller droplets.

13.1.2 Chemical Digestion

Chemical digestion occurs when the chemical makeup of dietary molecules is changed. Starch is changed into the disaccharide maltose in the oral cavity. Later, in the small intestine, all disaccharides are converted into monosaccharides such as glucose. Throughout the digestive tract organic catalysts, called enzymes, accelerate these chemical changes. In the oral cavity, for example, the enzyme amylase increases the rate by which starch is converted into the sugar maltose.

13.2 Anatomy – Digestive Tract and Accessory Structures

Most of the anatomy of the digestive system consists of a series of continuous chambers found on the ventral side of the body. They belong to a digestive tract which is over 30 feet in length. Food passes through the following series of chambers of this tract: **oral cavity** - **pharynx** - **esophagus** - **stomach** - **small intestine** - **large intestine** - **rectum.**

Several accessory structures also belong to the digestive system. These structures are either found in the chambers, such as the **salivary glands** and **teeth,** or structures that connect to the tract. The **liver, gallbladder,** and **pancreas** all connect by ducts to the duodenum, the first segment of the small intestine.

Oral Cavity - The oral cavity contains 32 permanent (secondary) teeth. There are four types based on their unique shape and size. The anterior **incisors** have sharp edges for cutting. The **canine** teeth have some ability to tear food. The **premolars** and **molars** have broad, flat crowns for crushing and grinding food. In any one of the four quadrant (quarters) of the mouth, their are eight teeth: 2 incisors (central and lateral), 1 canine (lateral and posterior to the lateral incisor), 2 premolars (posterior to the canine), and 3 molars (posterior to the premolars). The third molar (most posterior) is the wisdom tooth.

A tooth has three layers: the outer **enamel**, the **dentin** (middle layer) and inner **pulp**. The white enamel is the hardest substance in the body. The pulp contains the blood vessels and nerves.

The **crown** is the portion of the tooth, covered with enamel, above the line of the gingiva (gum). The **neck** is the constricted portion normally covered by the gum. The **root** is the part anchoring the tooth in the cancellous bone of either the maxilla or the mandible.

Several pairs of **salivary glands** secrete saliva through ducts into the oral cavity. The saliva contains amylase and moistens the food. Each **parotid gland** is embedded in the cheek, anterior and somewhat inferior to the ear. Each **submandibular gland** is near the posterior corner of the mandible. Each **sublingual gland** is under the mucous membrane lining the bottom of the oral cavity.

Skeletal muscles in the **tongue** contract to manipulate the food into a ball, or bolus, for swallowing. The surface of the tongue also contains the receptors for the sense of taste.

Pharynx - Food passes into the oropharynx through an opening from the oral cavity, the fauces.

Esophagus - The esophagus is a ten-inch muscular tube that accepts food from the pharynx. Peristalsis, rhythmic contractions of the tract wall, squeezes food along to the stomach.

Stomach - The **fundus** is the superior cap of the stomach. The **body** is the thick main portion. The **pylorus** is the constricted portion connected to the small intestine. The **cardiac sphincter** is a circular arrangement of smooth muscle fibers where the esophagus meets the stomach. The **pyloric sphincter** at the other end of the stomach has a similar makeup. These sphincters are controlled by reflexes to contract at certain times, preventing the backflow of food.

The stomach is a reservoir storing food, passing it on in small increments to the small intestine. The chemical digestion of proteins begins in this chamber. The enzyme pepsin, working in an acid environment (pH of 1 to 3), speeds up this digestion. HCl secreted into the stomach accounts for this acidity. Peristaltic action of the smooth muscle in the stomach wall contributes to physical digestion.

The stomach has four layers, which are also found in the next several chambers of the digestive tract. From the inside they are the **mucosa** (mucous membrane), **submucosa** (containing blood vessels), **muscularis** (smooth muscle), and **serosa**.

Small Intestine - The small intestine consists of three regions. The **duodenum** is the first 10 to 12 inches. This C–shaped region is followed by the **jejunum** (8 feet). The last 12 feet is the **ileum.** The ileum is separated from the large intestine by the **ileocecal valve**.

The mucosa of the small intestine has **villi** (sing., villus), finger-like projections that greatly increase the surface area of this lining. The small intestine is the main site of **absorption**, the passage of nutrient molecules from the digestive tract into the bloodstream. The chemical digestion of carbohydrates and proteins is completed here. The chemical digestion of lipids is started and completed here.

The stomach and small intestine are connected to **mesenteries,** sheetlike folds of serous membranes. Mesenteries are a continuation of the serous lining of the abdominal cavity, the **peritoneum.** They store fat and also contain blood vessels and nerves that supply the digestive tract.

Several accessory structure secrete substances into the duodenum.

The **pancreas** ranges from the C-shaped recess of the duodenum to along the bottom of the stomach. It secretes many enzymes, through a **pancreatic duct**, into the duodenum. It also secretes bicarbonate ions into this chamber. These ions buffer the acidic **chyme**, a mixture of partially-digested food passing into the small intestine from the stomach. The buffering action neutralizes this acidity, making the small intestine slightly basic (pH of 8 to 9).

The **liver** consists of two lobes. The larger right lobe is across from the stomach. The liver has many metabolic functions, such as converting glucose into stored glycogen and detoxifying poisons. It also manufactures bile, a substance stored in the **gallbladder**. This bulblike structure is under the right lobe of the liver. Bile, when secreted into the small intestine, emulsifies fats globules.

The gallbladder is drained by the **cystic duct**. The liver is drained by the **hepatic duct**. These two ducts merge to form the **common bile duct**, which connects to the duodenum.

Large Intestine - The large intestine, or colon, is 5 to 8 feet long. Its regions are the:

cecum - This is a pouchlike region where the ileum attaches. A thin appendix extends from it.

ascending colon - This rises vertically in the abdomen from the crest of the ilium on the right. It contacts the liver.

transverse colon - It meets the ascending colon at the hepatic flexure (bend). The transverse colon ranges across the liver and stomach, contacting the spleen on the left.

descending colon - It meets the transverse colon at the splenic flexure. This region of the large intestine extends from the spleen to the left iliac crest.

sigmoid colon - It is S-shaped, entering the pelvic cavity from the left iliac crest and ending medially.

The large intestine is the site of **reabsorption**. By this process most of the fluids from secretions added internally to the tract (e.g., saliva, gastric juice from the stomach) are reclaimed into the bloodstream. Whatever is not reabsorbed is **eliminated** from the body. The remaining feces, solid wastes, consists of some fluid, undigested food molecules, dead cells, and bacteria that normally inhabit this lower portion of the tract. This mass passes through the rectum when eliminated.

Rectum - The rectum is the last 7 to 8 inches of the tract. Solid wastes pass through it and leave through the **anus**, the opening at the end of the tract.

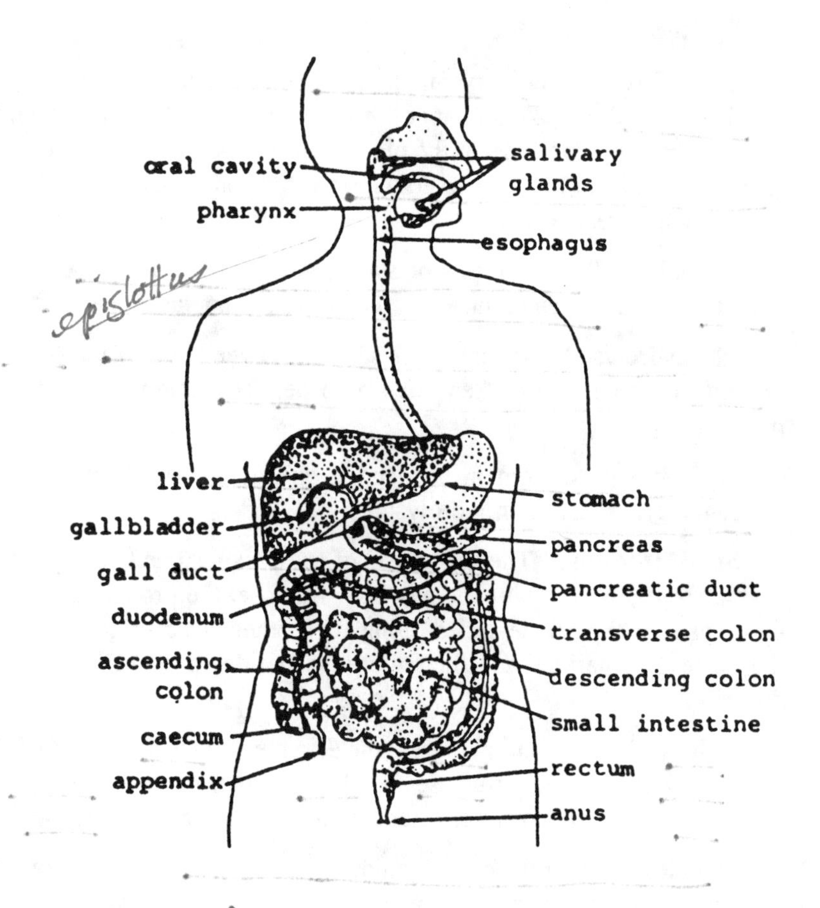

Figure 13.1 Anatomy – Digestive System

13.3 Physiology of the Digestive System

Food is digested through physical and chemical changes. Hydrolysis is chemical digestion. By the insertion of a water molecule, a covalent bond can be broken and larger molecules are reduced to smaller subunits. This occurs in three chambers of the tract. They are the oral cavity, stomach, and small intestine.

13.3.1 Hydrolysis

Oral Cavity - Salivary amylase accelerates the hydrolysis of starch into maltose. This change only begins here. Amylase is an enzyme that functions best at the pH of 7 in this cavity. The hydrolysis of proteins and lipids does not begin yet, as enzymes are lacking here for their breakdown.

Stomach - The hydrolysis of starch, swallowed with amylase, is inhibited. The acidic environment of the stomach is not favorable for it.

The hydrolysis of proteins begins here, catalyzed by pepsin. Large protein molecules are broken down into peptides, fragments of the proteins.

Lipid molecules remain unchanged chemically, as there aren't enough enzymes for their hydrolysis.

Small Intestine - The hydrolysis of starch is resumed, catalyzed by the secretion of pancreatic amylase. Maltose, the product of this hydrolysis, is further degraded along with other disacharides in the diet. The disacharides and their monosaccharides produced by hydrolysis are:

maltose digested into glucose—the enzyme is maltase
sucrose digested into fructose and glucose—the enzyme is sucrase
lactose digested into galactose and glucose—the enzyme is lactase

Peptides are hydrolyzed into amino acids, step by step. Enzymes such as trypsin and chymotrypsin speed up this process. Amino acids are the subunits produced.

After the emulsification of fat globules by bile, fat molecules (triglycerides) are broken down into fatty acids and glycerol.

13.3.2 Motility/Secretion

Motility is the movement of the food mass through the tract, created by the contraction of the smooth muscle layer in the wall of the tract. By **secretion**, substances are introduced into the tract to prepare and digest the food. Several different hormones control these events.

Gastrin - It is secreted by endocrine cells in the wall of the stomach. The stomach is also the target organ. The secretion of gastrin is enhanced by the arrival of protein molecules from the diet. The hormone increases gastric (stomach) motility and the secretion of pepsin and HCI. These responses facilitate the digestion of proteins.

Enterogastrone - This hormone is secreted by endocrine cells in the wall of the small intestine when fat molecules arrive in this chamber. Enterogastrone signals the stomach to decrease motility and secretion. This is an appropriate response, as fat molecules require time to be processed and digested in the small intestine.

Secretin - Secreted by the small intestine, it signals the pancreas when food molecules arrive in this chamber. The pancreas responds with some enzymes and bicarbonate ions. The ions buffer the acidity of the arriving chyme, the partially digested food mass.

Cholecystokinin-pancreozymin (CCP) - Both of these hormones are secreted from the small intestine as food arrives. Cholecystokinin signals the gallbladder to secrete more bile. Pancreozymin signals the pancreas to secrete more enzymes. All of these responses prepare and digest the food molecules.

13.3.3 Absorption/Reabsorption/Excretion

The final products of hydrolysis in the small intestine are monosaccharides, amino acids, fatty acids, and glycerol. In addition, vitamins (e.g., A or C) and minerals (e.g., Na or Ca) move through the tract unaltered. These substances are absorbed through the mucosa, entering the capillaries of the submucosa. Most of these substances are transported by the hepatic portal system to the liver for further metabolism.

Reabsorption rates in the large intestine are high. If, for example, 9,000 ml of fluid are secreted throughout the tract over 24 hours, 8,900 ml can be reabsorbed by the body.

The remaining, unabsorbed material in the colon is eliminated from the body, under the control of various reflexes (gastrocolic reflex).

CHAPTER 14

The Urinary System

14.1 Anatomy

The urinary system consists of two **kidneys**, two **ureters**, one **urinary bladder**, and one **urethra**. Each kidney consists of over one million microscopic units called **nephrons**. As the kidneys receive about 20% of the cardiac output every minute, the collective action of the nephrons controls the composition of the extracellular fluid (ECF). Therefore, the kidneys make a major contribution to body chemistry and homeostasis. As they regulate the composition of the ECF, they produce urine.

14.1.1 Kidney

Each kidney is slightly above the waistline. The location of the kidney is retroperitoneal, meaning that it is posterior to the peritoneum, a serous membrane lining the abdominal cavity. Each kidney is between the peritoneum and the skeletal muscles of the dorsal body wall.

The size of the kidney is about 5 inches by 2 inches by 1 inch (length, width, thickness). A **renal artery** and **renal vein** enter it at its medial notch, the **hilum**.

Each kidney is bound by a thin, leathery **capsule**. Internally, this organ is composed of two layers. The outer **cortex** is reddish, with an abundant supply of blood vessels. The inner **medulla** consists of about one dozen **renal pyramids**. These wedges have a striated appearance.

Although the cortical tissue is mainly external, some of it does penetrate deeper into the kidney as **renal columns**. These columns alternate with the pyramids. The pointed ends of the pyramids, the **papillae,** project deep and medially into the kidney. Each papilla fits into a funnel-like structure, the **calyx**. All of these calyces converge into a deep, medial cavity called the renal **pelvis.**

14.1.2 Other Organs

The **ureter** is a tube, 10 to 12 inches long, that extends from the pelvis of the kidney. This muscular tube propels the urine into the **urinary bladder** by peristalsis. The bladder, located in the pelvic cavity, stores the urine.

When the bladder becomes filled with several hundred ml of urine, a reflex (micturition reflex) leads to a contraction of the smooth muscle layer in its wall. This forces urine out of the bladder and through the **urethra.** This is the final passageway of the urinary tract, which is inferior to the bladder. If two sphincter muscles in the urethra relax as the bladder contracts, urine is eliminated from the bladder.

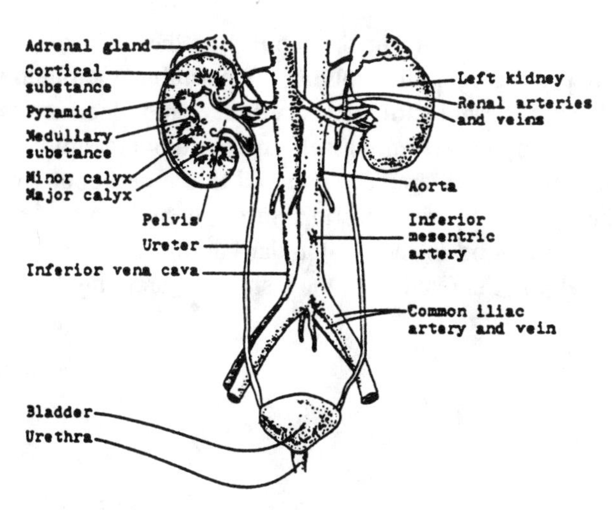

Figure 14.1 Gross Anatomy of the Kidney

14.1.3 Nephron

The nephron is made up of a **renal corpuscle** plus several regions of **tubules.**

The renal corpuscle of this microscopic unit consists of the **glomerulus,** a ball of capillaries, and a double-walled cup, the **Bowman's capsule**. The renal corpuscle is the site of **filtration,** the passage of plasma substances from the glomerulus into the Bowman's capsule.

From the renal artery, a series of arteries and smaller vessels send blood into each glomerulus. An **afferent arteriole** sends blood into the glomerulus directly. An **efferent arteriole** transports blood away from the glomerulus. The efferent vessel breaks down into a second set of capillaries, the **peritubular capillaries.** They surround the tubular portion of the nephron.

The tubular portion of the nephron has four main regions. Filtrate from the Bowman's capsule first passes into the **proximal convoluted tubule** in the cortex. Next, the filtrate enters the **loop of Henle,** first through its descending limb and then through its ascending limb. This loop extends deep into the medulla. From there, fluid enters the **distal convoluted tubule**, also in the cortex. From that tubule, fluid passes into the **collecting duct**, which passes deep into the medulla.

After filtration, fluid in the tubules of the nephrons undergoes two more processes, both involving the peritubular capillaries: **tubular reabsorption** and **tubular secretion**. Some blood is not filtered and passes into the efferent vessels and peritubular capillaries. Many substances that are filtered are returned to the peritubular capillaries from the tubules by reabsorption, often at high rates (e.g., water, glucose). The blood in the peritubular capillaries collects into other vessels that join the renal vein, the last vessel draining the organ.

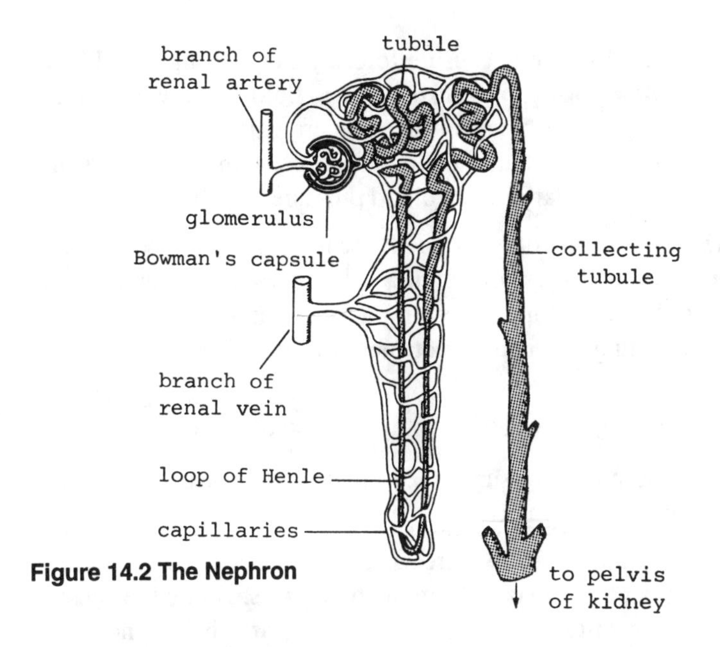

Figure 14.2 The Nephron

14.2 Renal Physiology

The nephrons carry out three processes that regulate the composition of the extracellular fluid and form the urine: **filtration**, **reabsorption**, and **secretion**.

14.2.1 Filtration

Filtration is the first step of renal physiology. By filtration substances leave the blood plasma, passing from the glomerulus into the Bowman's capsule. Large amounts of most substances (e.g., water, glucose, sodium) are filtered. The one exception is the plasma proteins, as these molecules are too large to pass through the pores in the walls of the glomerulus. Therefore they do not leave the blood.

The filtrate is produced by an interaction of several pressures. **Blood pressure** is the major one, tending to force substances from the capillary into the capsule. For a capillary it is abnormally high at the glomerulus. One example is 70 mm of mercury.

Two pressures oppose the blood pressure, as they are responsible for the movement of substances in the opposite direction. A **capsular**

pressure in the Bowman's capsule forces fluid back into the blood. By a **colloidal osmotic pressure** (COP), fluid is drawn into the blood by osmosis. The blood in the glomerulus remains hypertonic because the plasma proteins, colloids based on their size, are not forced out of the capillary by the blood pressure.

If the capsular pressure and COP are 20 and 30 respectively, as examples, the blood pressure forcing substances out of the blood is still greater. The difference between the blood pressure, and the two pressures opposing it, yields a net pressure, the **effective filtration pressure** (EFP):

$$EFP = 70 - (20 + 30) = 20$$

In this example a net pressure of 20 units forces plasma substances into the blood of the glomerulus.

Filtration is a nonselective process, as all plasma substances (except the plasma proteins) pass from the blood plasma into the capsule in significant quantities. The cells of the blood are also not normally filtered.

Filtered substances next pass from the Bowman's capsule into the proximal convoluted tubule of the nephron.

14.2.2 Reabsorption

Most filtered substances are returned to the blood by the second step of renal function, reabsorption. Specifically, substances pass from the tubules of the nephron into the blood of the peritubular capillaries.

Usually reabsorption occurs at high percentages rates. For example, if 190 liters of water are filtered over 24 hours, around 189 liters of water can be reabsorbed. Therefore, the rate of reabsorption for water is over 99%. Normally the reabsorption rate of glucose is 100%. If 300 grams of glucose are filtered over 24 hours, all 300 grams are usually reabsorbed.

If a substance is not secreted, a possible last step of renal physiology, the amount of a substance eliminated is equal to the amount filtered minus the amount reabsorbed. The amount that is not reabsorbed does not reenter the blood. It will remain in the tubules of the nephron and be eliminated. As one example:

190 liters of water filtered – 189 liters reabsorbed = 1 liter eliminated

Unlike filtration, reabsorption is a very selective process. Substances are returned to the blood to the degree they are needed for body metabolism. To fulfill this need, reabsorption is controlled.

Sodium reabsorption is controlled by **aldosterone**, a hormone secreted by the nearby adrenal cortex. If more sodium is needed, more aldosterone is secreted. It signals the nephron tubules, particularly cells in the wall of the proximal convoluted tubules, to transport more sodium back into the blood. If less sodium is needed by the body, less aldosterone is secreted to stimulate sodium reabsorption.

The reabsorption of water depends on the reabsorption of sodium at the proximal convoluted tubule. The reabsorption of water there is **obligatory**. This means that it must occur. The blood in the peritubular capillaries becomes hypertonic as it accumulates reabsorbed sodium. The tubular wall is permeable to water which follows sodium ions by osmosis. This accounts for a large amount of the total reabsorption of water.

The reabsorption of water also affects blood pressure. Water reabsorption adds more fluid to the vessels. Just as more water in a garden hose increases its pressure, adding more fluid to blood vessels increases their pressure. Cells in the junxtaglomerular apparatus, cells near the afferent vessel, detect blood pressure changes. These cells secrete **renin**, an enzyme that converts the plasma protein angiotensiongen into **angiotensin**.

Angiotensin signals the adrenal cortex to secrete aldosterone which stimulates sodium and water reabsorption. One example of this chain of events occurs to oppose blood pressure if it decreases:

blood pressure decreases - renin increases - angiotensin increases - aldosterone increases - sodium reabsorption increases - water reabsorption increases - blood pressure increases

In the distal convoluted tubules and collecting ducts, the reabsorption of water is **facultative**, meaning it varies and is not obligatory. A small amount of water can be reabsorbed near the end of the nephron, depending on signaling from the hormone **ADH** (antidiuretic hormone). If more water is needed, more ADH is secreted from the posterior lobe of the pituitary. Therefore, on the output side, more water is reabsorbed and less is eliminated in the urine.

The **hypothalamus** senses the solute concentration of the blood and signals the posterior lobe to make this final adjustment of reabsorption. The hypothalamus also has a thirst center that controls water intake by drinking. If the blood is too hypertonic, one example that restores a balance is:

hypertonic blood - more water needed in ECF (blood) to dilute the solutes in the blood - hypothalamus senses this condition - more ADH is secreted - more water is reabsorbed and less is eliminated. In addition, the sense of thirst increases and can add water on the input side.

Cells composing the walls of the nephron are specialized to carry out unique functions. The cells of the proximal convoluted tubule have microvilli, establishing a brush border. This increases their surface area for reabsorption.

Cells in the ascending limb in the loop of Henle have numerous mitochondria. These organelles, or powerplants, give the ascending cells the energy to pump sodium back into the fluid surrounding the descending limb. This makes the surrounding fluid more hypertonic. This leads to the greater reabsorption of water at it leaves the tubule by osmosis when passing through the wall of descending limb.

Most filtered substances have an upper limit, or threshold, over which they cannot be reabsorbed. Glucose, for example, normally has a 100% reabsorption rate. However, at concentrations over 180 mg/100 ml (milligrams per 100 milliliters) in the filtrate, the tubule cells lack the time or energy to return the extra levels of glucose into the blood.

In the blood and filtrate of a diabetic (diabetes mellitus), the blood sugar level can be very high (i.e., 1,200 mg/100 ml). The extra glucose that cannot be reabsorbed will be eliminated in the urine. A normal level of sugar in the blood and filtrate (i.e., 100 mg/100 ml), can be controlled by normal renal functions and completely reabsorbed.

14.2.3 Secretion

Some substances undergo a third process called secretion. By this process, substances are transported from the blood in the peritubular capillaries into the nephrons near their ends—the distal convoluted tubules and collecting ducts. Potassium ions, which are usually reabsorbed at a rate of 100%, may be added back to the nephron by this third step. Hydrogen ions are often secreted to eliminate them from the blood. This opposes their buildup in the blood, a condition called acidosis.

The secretion of hydrogen supplements buffer systems in the blood. Buffers are chemicals that react with free hydrogen ions in the blood and eliminate them by making them parts of other compounds. For example, the buffer compound sodium bicarbonate reacts with free hydrogen ions to produce this effect.

If a molecule such as urea (nitrogen waste product) is filtered and not reabsorbed, it will be eliminated from the body by passing through the following structures:

renal artery - other arteries - afferent arteriole - glomerulus - proximal convoluted tubule - loop of Henle (descending limb) - loop of Henle (ascending limb) - distal convoluted tubule - collecting duct - renal papilla - calyx – pelvis - ureter - bladder - urethra

Numerous collecting ducts converge at the papilla end of each pyramid of the medulla. From here, the funnel-like calyces accept molecules that will enter the larger structures of the urinary system for elimination.

CHAPTER 15

The Reproductive System

The male and female reproductive systems produce sex cells. In addition, the female system provides the internal environment for fertilization and for the development of the embryo and fetus.

15.1 Male Reproductive System

The male reproductive system consists of the pair of testes (sing., testis), ducts, glands, and external genitalia.

Testis - Each testis, an oval-shaped structure, is suspended in the sac-like **scrotum**. The testis is encased by a fibrous capsule, the **tunica albuginea**. The dense fibrous tissue of this capsule continues internally as a series of **septa** or walls. These walls subdivide the interior of each testis into a number of **lobules** or chambers. Each lobule contains **seminiferous tubules** and **interstitial cells** (**cells of Leydig**).

The seminiferous tubules are the site of **spermatogenesis**, sperm cell production. The tubules from each lobule converge and join posteriorly at the **rete testis**. The rete testis connects to the **epididymis**, a long structure next to each testis posteriorly. The **vas deferens**, a major duct of the male reproductive tract, exits from the base of each epididymis.

The interstitial cells produce and secrete the hormone **testosterone**. This substance promotes the development of the male reproductive organs and secondary male characteristics (e.g., pitch of voice, hair growth on the face).

Ducts - The ducts form a connected series of passageways for the migration of sperm cells when they are ejaculated. This series of ducts forms the male reproductive tract.

The **epididymis** is a site where sex cells are stored and where they mature. Each **vas deferens (ductus deferens)**, along with blood vessels and nerves, is part of a **spermatic cord**. From each side of the scrotum, this combination of structures passes through the **inguinal canal**, a passageway leading from the scrotum into the pelvic cavity. The vas deferens conducts ejaculated sperm cells into this cavity.

From its ascent into the pelvic cavity, each vas deferens approaches the bladder from the anterior side. It passes over the top and descends on the posterior side, medial to where each ureter enters the bladder. The expanded end of the vas deferens is the ampulla. The two ampullae join, forming the **common ejaculatory duct**. This single structure passes through the **prostate gland**, which is inferior to the bladder. Within this gland the common ejaculatory duct joins the urethra.

The urethra has three regions. The **prostatic urethra** is the first region, exiting from the bladder. From the point where the prostatic urethra merges with the common ejaculatory duct, the urinary and reproductive systems of the male are served by a common tract. The **membranous urethra** is a short region after the prostatic urethra, but before the tract enters the penis. The **penile urethra** is the last region, passing through the cavernous (spacious) tissue of the penis.

Glands - Several glands add seminal fluid to sperm cells migrating through the male tract. Semen is the mixture of sperm cells and seminal fluid formed by this addition.

The **epididymis** is a part of the tract, storing cells from the testis before they enter the vas deferens. It also adds a small percentage of the seminal fluid.

A pair of **seminal vesicles** add about 60% of the seminal fluid. Each gland secretes fluid into the ampulla of the vas deferens. The single, large **prostate gland** adds about 30% of the seminal fluid to the prostatic urethra. A pair of **Cowper's (bulbourethral**) glands add a small amout of fluid to the urethra as it passes into the posterior part of the penis.

External Genitalia - The **scrotum** is covered with skin externally. This structure is made up of two sacs separated by a septum. Each sac houses one testis, epididymis, and the beginning of each spermatic cord.

The **penis** contains three masses of cavernous tissue. This tissue has spaces that can fill with blood, causing the penis to become erect. More blood enters the penis when arteries serving this organ dilate, producing an erection. The two large **corpora cavernosa** (sing., cavernosum) are separated by a septum. The **corpus spongiosum** surrounds the urethra. It expands into the **glans,** the blunt and enlarged end of the penis.

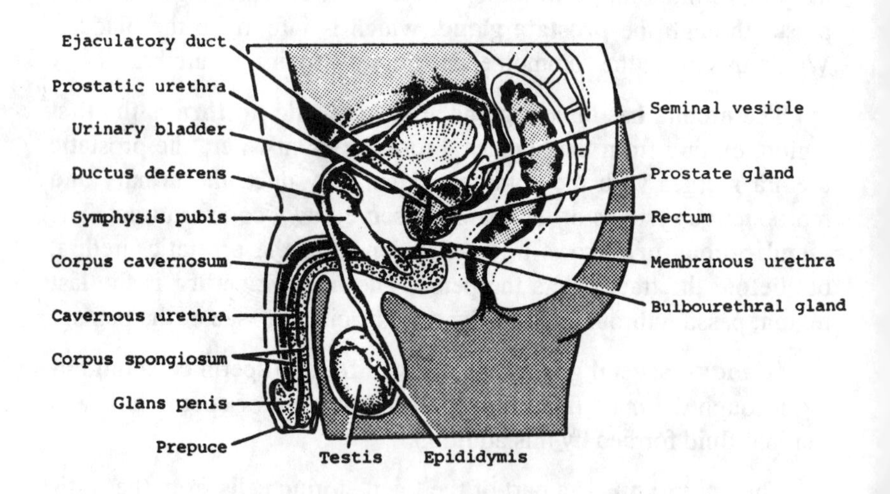

Figure 15.1 Male Reproductive System – Sagittal Section

15.2 Female Reproductive System

The female reproductive system consists of a pair of ovaries and two uterine tubes, plus the uterus and vagina. The external genitalia are also part of this system.

Ovary - Each ovary is attached to the broad ligament, a fan-shaped structure that is posterior to the uterus and oviducts. The pair of ovaries store nearly one-half million immature sex cells. During ovulation one sex cell matures and is released from the ovary into the oviduct. The ovaries also secrete changing levels of two hormones, **estrogen** and **progesterone** during each female reproductive cycle.

Oviduct - On each side of the body, the expanded end of the oviduct (**uterine tube**, **Fallopian tube**) curves over the ovary. At its opposite end the oviduct enters the fundus, the superior region of the uterus. A sex cell released from the ovary during ovulation enters the oviduct, drawn in by the muscular contractions of this part of female reproductive tract. Normally the cell is fertilized in the oviduct, becoming an embryo of about 200 cells by the time it arrives in the cavity of the uterus.

Uterus - The uterus resembles an inverted triangle. The **fundus** is the dome-shaped region. The **body** is inferior to the fundus, tapering into a **cervix**. The cervix is the most inferior region, which fits into the vagina.

The uterus has three layers. The **endometrium** is a mucous membrane that lines the cavity of this organ. It is the site where the embryo implants after arriving from the uterine tube. It develops a vascular and glandular makeup to receive and maintain this early stage of development. The **myometrium** is the middle layer of smooth muscle. Its contraction during labor expels the fetus during parturition (birth). The **epimetrium** is a serous membrane that lines most of the external surface.

The uterus provides the internal environment for the prenatal development of the embryo and fetus.

Vagina - The vagina is the organ of copulation, a four- to six-inch muscular tube that accepts the penis. Bartholin glands at its external opening secrete substances that serve as lubricants for this process.

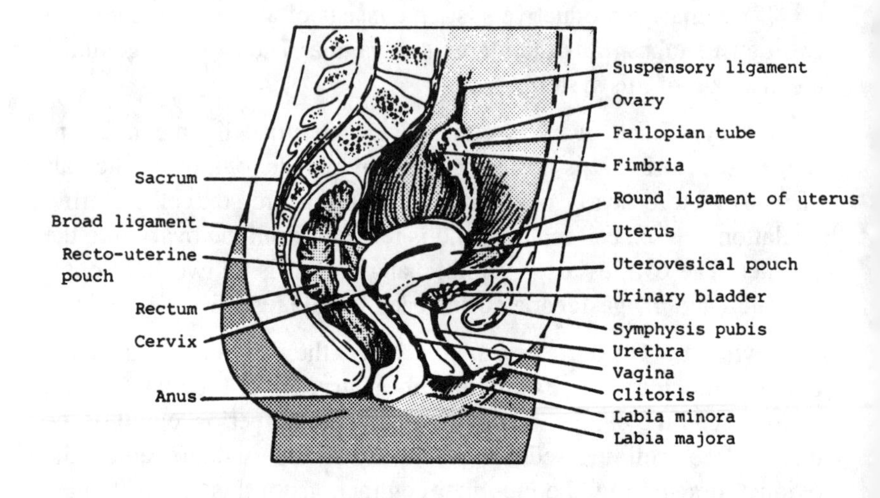

Figure 15.2 Female Reproductive System – Sagittal Section

External Genitalia - A pair of **labia majora** are external folds that are lateral to the pair of medial **labia minora**. These medial folds unite anteriorly to form a small mass of erectile tissue, the **clitoris**. Posteriorly these folds unite to form the **posterior fourchette**. Within the labia minora, the opening of the urethra is anterior to the opening of the vagina.

The **vestibule** is the region extending from the clitoris to the posterior fourchette. The **mon pubis** is an elevated region that is anterior to the vestibule.

15.3 Female Reproductive Cycle

The events of the female reproductive cycle are controlled by the interaction of many hormones: FSH, LH, estrogen, and progesterone. Over a 28-day cycle, used here as an example, the major events are:

1. During **menstruation** there is a disintegration and loss of the endometrium. This occurs if an early embryo does not enter

the uterus and implant to the inner lining. In this case the vascular and glandular thickening of the endometrium is not necessary to accept and maintain the embryo.

The concentration of all hormones is low through this period. Menstruation can last over the first five days.

2. The secretion of **FSH**, from the **anterior lobe** of the **pituitary gland**, increases early in the cycle. FSH eventually stimulates the development of an **ovarian follicle**, a layer of cells surrounding one female sex cell in one of the ovaries. This event signals the onset of the **follicular phase**, the stage of the cycle after menstruation.
3. Cells of the active ovarian follicle secrete increased amounts of **estrogen**.
4. The surge in estrogen stimulates a proliferation of the endometrial lining, causing it to thicken.
5. By negative feedback, the increase of estrogen produces a gradual decrease in FSH, probably by signaling the hypothalamus that, in turn, controls the anterior lobe.
6. As FSH is gradually decreasing, the secretion of **LH** is gradually increasing. Peak levels of LH produce ovulation. This event occurs at about 14 days prior to the beginning of the next cycle. In a 28-day cycle, this is the midpoint. At this midpoint, the follicle is converted to the **corpus luteum**.
7. The conversion to the corpus luteurn signals the beginning of the **luteal phase**. This stage takes up about the last half of the entire cycle. Initially in this phase the corpus luteum is maintained by LH.
8. The corpus luteum secretes increased levels of **progesterone** immediately after ovulation.
9. Increased progesterone, plus a new surge in estrogen secretion several days into the luteal phase, produces further maintenance and development of the endometrium.

10. The increase in progesterone and estrogen also feeds back to the hypothalamus to diminish the level of LH as the luteal phase progresses.

11. As LH drops off through the luteal phase, the corpus luteum cannot be maintained. It gradually degenerates, leading to a decrease in estrogen and progesterone during the final days of the luteal phase.

12. As estrogen and progesterone drop off by the end of the cycle, the endometrium is no longer maintained. It begins to disintegrate, leading to menstruation and the onset of the next cycle.

If fertilization and implantation occur, the endometrium does not degenerate. When implantation occurs, the woman becomes pregnant. Estrogen and progesterone levels remain high, supporting the development of the embryo and fetus. The production of an additional hormone, the **human chorionic gonadotropin** (HCG) also supports the development of the embryo and fetus. Some of this hormone is excreted in the urine. Its detection by testing a urine sample is the basis for the pregnancy test.

CHAPTER 16

Development

16.1 Embryonic Development

Gestation, the period of pregnancy, normally takes place over about nine months. Development before birth (prenatal development) begins with **fertilization**, the union of the sperm cell and female sex cell. Fertilization produces the **zygote**, or fertilized egg, in the oviduct. A series of events unfolds producing the **embryo** (first two months) and **fetus** (last seven months).

The zygote begins to divide immediately to produce an early embryo by mitosis The early stages of the embryo are: 2-cell, 4-cell, 8-cell, 16-cell, and 32-cell. This early division pattern is called **cleavage**, as the cell number increases without any increase in the cytoplasm. The 32-cell stage, for example, is the same size as the zygote.

Several days after fertilization a mulberry-shaped mass, the **morula**, is formed in the oviduct. The morula develops into a **blastocyst**, a fluid-filled hollow ball of cells. This blastocyst will enter the uterus and implant on the endometrium at about 8 to 10 days after fertilization.

The early embryo changes from a hollow ball of cells through two processes: **morphogenesis** and **cell differentiation**. Morphogenesis is the movement of cells to establish a human outline. Cell differentiation means that cells specialize into different kinds. All cells descending from the zygote have a full complement of chromosomes and genes through mitosis. Depending on which genes are expressed in a group of cells, they specialize to become nerve cells, muscle cells, etc.

Morphogenesis and cell differentiation establish three primary germ layers in the embryo by about two weeks after fertilization: the **ectoderm**, **mesoderm**, and **endoderm**. A continuation of these two processes in these germ layers form all future tissues and organs. Some examples are:

ectoderm (outer layer) - epidermis of skin, tooth enamel, lens and cornea of the eye, nervous system

mesoderm (middle layer) - dermis of skin, connective tissues, skeleton, skeletal muscles, circulatory system

endoderm (inner layer) - digestive tract and accessory organs, respiratory tract, kidney nephrons, bladder, several endocrine glands

Other significant events of embryonic development include:

week 3 - The embryo is surrounded by a membrane, the **amnion**, and amniotic fluid. The **placenta** starts to form.

week 4 - The heart beats and is pumping blood. The limb buds form. The placenta is fully formed and working.

week 6 - The limb buds develop digits. A skeleton of cartilage forms.

week 8 - All internal organs are produced. The embryo appears humanlike.

16.2 Fetal Development

Refinements of morphogenesis and differentiation produce the following results as the fetus develops.

month 3 - The head is abnormally large when compared to the rest of the body. It is distinct with well-formed eyes and ears. The biological sex is identifiable by examining the external organs. Bone tissue gradually replaces the cartilage of the skeleton.

month 4 - A bony skeleton is established. The skeletal muscles contract and produce body movements. The head is not as disproportionately large.

month 5 - The internal organs continue to develop. The fetus may reach a weight of 3⁄4 pound or 1 pound.

month 6 - Fused eyelids begin to open. The fetus may reach a weight of 1.5 pounds.

months 7-9 - There is a tremendous weight gain, with the fetus reaching 6 to 8 pounds normally by birth. The body length normally reaches 18-22 inches. The **fontanels**, soft membranous patches between the cranial plates, are distinct.

16.3 Parturition

Parturition is the process of birth. The fetus is expelled from the uterus and through the vagina by the contraction of the myometrium.

Oxytocin, a hormone secreted from the posterior lobe of the pituitary gland, signals the uterus to develop the forceful contractions for birth. **Labor** occurs when the uterus contracts approximately every 15 minutes. The length of each contraction during labor is over 30 seconds. The cervix dilates as the baby is born. The amnion may rupture, releasing the amniotic fluid through the vagina.

Shortly after the delivery of the baby, the placenta (afterbirth) passes through the vagina, pushed by uterine contractions.

16.4 Postnatal Development

Development after birth, postnatal development, progresses over a series of stages: **infant**, **child**, **adolescent**, and **adult**.

infant - For the first 24 months after birth, the infant develops many sensory and motor abilities. These abilities include focusing on objects as well as grasping and walking. Patterns of speech and language development unfold.

child - Many emotional characteristics are established during this period. Physical growth is substantial. Refinements of mental capacities, such as reading, occur.

adolescent - Beginning with the onset of puberty, this period begins with a substantial growth spurt. Behavioral changes occur as the individual begins to imitate adulthood. Many physiological changes occur with attainment of full adult stature, usually by the late teens.

adult - Reached by about the age of 20, one's physical peak is usually attained in the early years of adulthood. Muscular strength and the ability to respond homeostatically, as examples, are optimal. The onset of aging is marked by a decrease in the number of functional cells in all organ systems. The rate of this decline, however, is extremely variable. Patterns of exercise and diet greatly influence it.